MOTHER LODE MEMORIES

MOTHER LODE MEMORIES

A Pictorial History

by Dr. R. Coke Wood and Leonard Covello

Fresno   California

A Division of Book Publishers, Inc.

1979

Copyright © 1979
Valley Publishers
All rights reserved. No part of this book may be reproduced in
any form without the written consent of the publishers.
Manufactured in the United States of America.
Valley Publishers
8 East Olive Avenue
Fresno, California 93728

Library of Congress No. 79-67345
ISBN 0-913548-70-7

Contents

I	Mother Lode Rivers	1
II	El Dorado County	5

Lumber Industry In El Dorado County: 27

III	Amador Area	37
IV	The Calaveras Region	65

Black Bart: 101

"Clamper Wall" In Murphys: 104

V	Tourist Attractions of Calaveras	107
VI	Tuolumne County	121
VII	Rivers of Tuolumne	157
VIII	Mariposa	167

Bands, Entertainment
And Special Events
In The Mother Lode Country: 195

Bibliography: 204

Photographic Credits: 204

Index: 205

Publisher's Preface

This book does not attempt to tell all the history of the gold country in California and the gold rush to that area. Its discussion and pictures are limited to what is called the "Mother Lode," an area from Mariposa on the south to Georgetown on the north, a distance of about 120 miles along the foothills of the Sierra Nevada Mountains. It includes five complete counties (El Dorado, Mariposa, Amador, Calaveras and Tuolumne) and also takes in a portion of Stanislaus County.

The authors have sought out important people of the times, and places and events of special interest, including gold rush hotels, bars, gambling houses, churches and fraternal organizations.

Their scope is not limited to gold mining, even though that started the rush to these camps and towns. They have included many different kinds of events and industries relevant to the area, even though in some cases coverage is limited to a photograph or two of the subject.

A short explanation about the use of the term "gold" is in order. If gold is defined as wealth, then resources which produce wealth can also be called gold. Hence lumber and farming are called "green gold," recreation and tourism are called "tourist gold," etc.

Yellow gold remains the starting point, however, for not only did the metal itself produce vast wealth, but these other resources were not developed until after the gold rush, even though many of them were abundant in the area much earlier.

Both authors have known the Mother Lode intimately. Photographer Leonard Covello is a former musician who played in most of the central Mother Lode dance halls many weekends during the 1920s and 1930s. In his youth he lived in Angels Camp with his uncle, who operated the Palace Barber Shop on Main Street.

Richard Coke Wood is an author, teacher and historian, and he operates the Old Timers Museum in Murphys. As a result of his extensive knowledge of California and the Mother Lode, he received the title "Mr. California" in a joint resolution of the California State Legislature in 1969.

It has been over 125 years since gold was discovered at Coloma. In that time the Mother Lode country has known romance, adventure, excitement, and vast struggles for riches. We are proud to make these *Mother Lode Memories* available to you.

Valley Publishers

Acknowledgements

It takes the cooperation of a great many people to put together a book like this. We have been fortunate to receive such cooperation from a number of individuals and organizations, and to them we express our sincere appreciation and gratitude.

We especially thank Ardath Covello for her patience and determined efforts in copying and correcting the longhand versions of the manuscript and photo captions.

We also thank Raymond Hillman, curator of history at the Pioneer Museum and Haggin Galleries in Stockton, for reading and proofing the manuscript. Mr. Hillman has been an instructor in California history at San Joaquin Delta College and is especially well informed on the history of the Mother Lode country.

Many people in each of the Mother Lode counties gave us both information and inspiration. We are grateful to the following:

El Dorado County: Bruce Robinson, Placerville; the El Dorado County Museum, Placerville; El Dorado County Chamber of Commerce, Placerville; Harriman Golden Trail Union School District, Gold Hill.

Amador County: Blazing Star Mine, West Point; D. G. "Duff" Chapman, Jackson; Mrs. Norma Cuneo and her aunt, Frances Schacht, Jackson; Mrs. Margarette Pease, Amador County Museum; D'Agostini Winery, Plymouth; Raymond Drew, Amador Chemical Company; Barbara Cunetto, Daffodil Hill; Sutter Creek Inn, Sutter Creek.

Calaveras County: Sam Bryan, Bender Forest Products Corporation; Willard Fuller, Calaveras Cement Division, Flintkote Corporation; *Las Calaveras,* publication of the Calaveras Historical Society; Mrs. Judy Cunningham, Calaveras Heritage

Council; Old Timers Museum, Murphys; Mrs. Marian Middlin, (daughter of Frank Tower, who owned the Royal Consolidated Mine), Murphys; Calaveras County Chamber of Commerce, Angels Camp.

Tuolumne County: Carlo M. De Ferrari, Sonora; Tuolumne County Historical Society; William Dyer, Columbia College; Frank McCormick, Tuolumne County Historical Society; Jim Oliver, Tuolumne Chamber of Commerce; William J. Lange, manager, East Bay Utility District, Lodi; F. F. Momyer, president, Pickering Lumber Company, Standard; West Side Lumber Company, Tuolumne.

Mariposa County: Mariposa County Historical Society; *Mariposa Gazette* and owners Margaret Campbell and Dexter Campbell; Alice Sargent and Marian Thompson, Mariposa County Historical Society; Allan Hague, Coulterville; Northern Mariposa-Coulterville-Pepper Museum.

If we have failed to mention the name of anyone who has helped, it is purely an oversight and does not reflect a lack of appreciation.

R. Coke Wood
Leonard Covello

September 1979

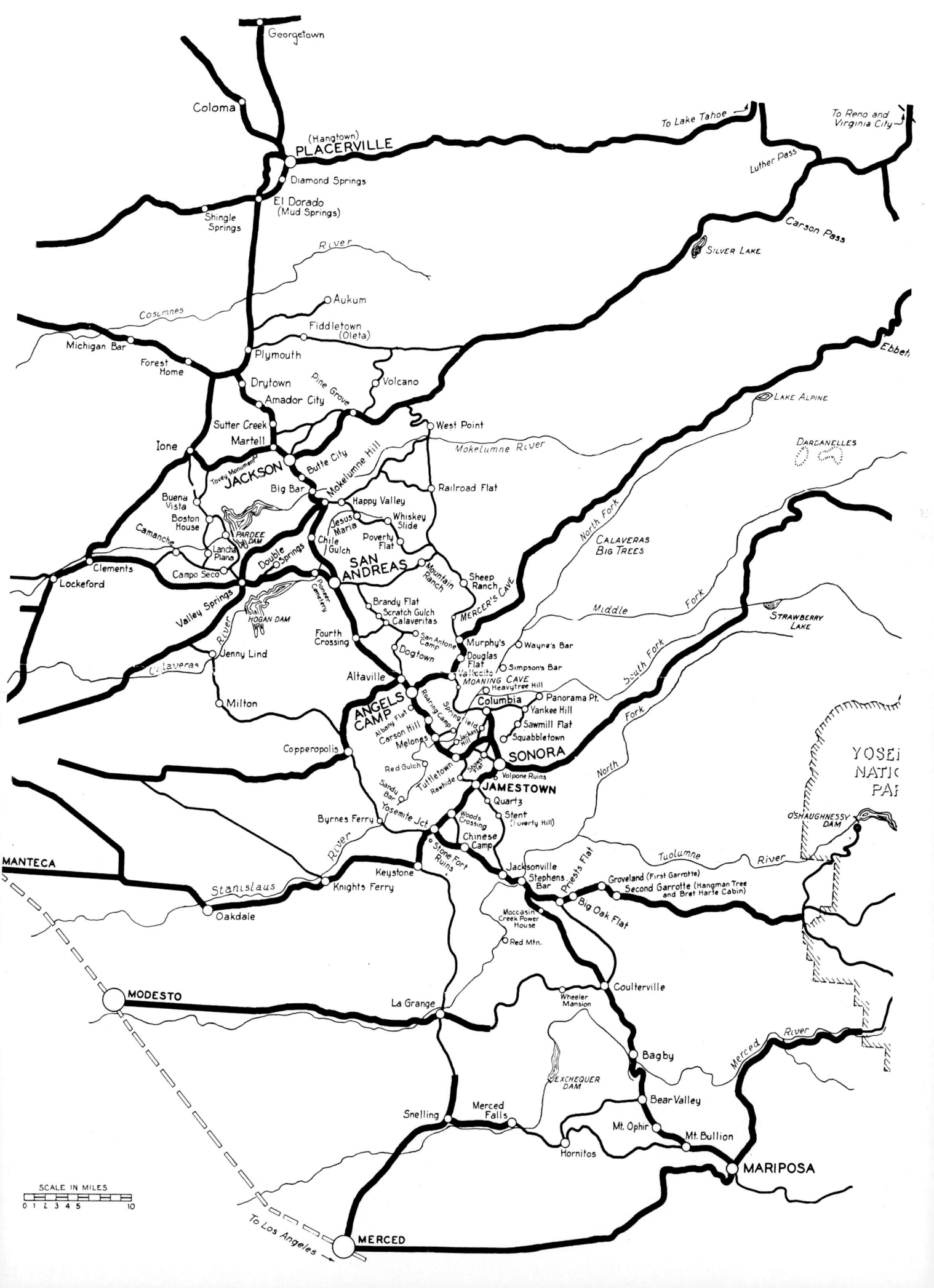

I

The Mother Lode Rivers

The placer or free gold on which the California Gold Rush was based centered around the rivers of the Mother Lode and gold country. It was in the gravel of the American River at Coloma, on January 24, 1848 that James Marshall picked up those little golden flakes that set off the search for gold among the sand and gravel of the rivers draining the Sierra Nevada mountains.

That was the beginning of an era, but only a moment in the age-old story of gold in this area. Here is that story as geologists have reconstructed it.

At one time, during the Paleozoic era, a great shallow sea covered the West, from what is now the Pacific shoreline to the mountains that were to become the Rockies. Great deposits of earth and debris were washed down over the centuries, and these deposits became the floor of the Jurassic Sea.

Ages later, internal pressures erupted and greatly changed the earth's surface. It was then that the Sierra Nevada mountains were heaved upward from the floor of the sea. The Coast Ranges were probably formed at the same time.

This upheaval was caused by terrific pressures which compressed the deposits on the floor of the Jurassic Sea into slate and schists.* At the same time, masses of liquid granite were forced upward into the slate and schist to form the bedrock that was lifted above the water to become the first Sierra Nevada mountains. Between these mountains and the Coast Ranges was an extensive gulf.

Those first sierra (saw-toothed) mountains had fairly even surfaces, but at once the weather began to wear down their slopes, and streams which were to become river channels and valleys

---

*A schist is a crystalline rock consisting of mineral ingredients so arranged as to impart a more or less laminar structure, that may be split easily into layers.

began flowing to the West. These streams are called Tertiary rivers. They are not identical to the present rivers because of a later upheaval which changed their courses. However, these meandering streams deposited debris and considerable amounts of gold in the gravel beds, and the gold was loosened from the bedrock by the erosive action of the weather and the water.

Miners and geologists have been able to trace the courses of several rivers of Neocene origin and have named them after the present-day rivers, formed after the second upheaval of the mountains. They can be identified as the Tertiary Yuba, American, Mokelumne, Calaveras and Tuolumne Rivers. It is apparent that these streams descended in a slow gradual course, and over the long period of their existence deposited great amounts of debris and gravel.

In a later period the ridges of the Sierra Nevada were heaved skyward by internal pressures until a mighty mountain range was formed, separating the interior of the continent from the Pacific Coast area. At the same time, the great internal heat and convulsions forced liquid rock in the form of lava out through the fault lines near the mountain summits. Much of this liquid rock flowed down and filled the canyons of the ancient rivers, sealing in them the deposits of gold.

After this upheaval the Sierra Nevada, with its sides covered by hardened lava, was a gigantic mountain range standing as high as 20,000 feet above sea level. In time the summits became covered with snow and ice, and these great glaciers began sliding down the mountains to form canyons and valleys. As the glaciers reached the lower altitudes, they melted into streams of water. These torrential Pleistocene rivers did not follow the course of the old Neocene rivers but cut new, more precipitous channels from the solid rock, cutting across the channels of the old river beds and forming the deep V-shaped channels of the present Feather, Yuba, American, Mokelumne, Stanislaus, Tuolumne and Merced Rivers, as well as some smaller ones.

The miners of the Gold Rush of 1848 owe a great deal to the process of nature that formed the Tertiary, or pre-glacial, rivers. Over the ages these rivers were piling up the free, or placer, gold in the gravel and sand bars along their routes. They deposited their debris in depressions, some narrow and deep, some broad and shallow. If they have not been too greatly eroded by more recent streams, these old channels may still be traced. Though their courses are irregular, their streams are full of gravel. There are many examples of this, but one of the most spectacular is the gravel remains of the prehistoric Yuba River at the Malakoff Diggings in northern Nevada County, twenty-five miles north of Nevada City, now a state park.

The flood plains on the lower levels of these old rivers are wide and extensive, but toward the summit of the Sierra, they become smaller and less distinct. The water of these prehistoric rivers washed out the gold implanted in the bedrock of the old Jurassic Sea, which had been forced upward by the upheavals. Sometimes this gold existed in large quantities, especially in the fissures of quartz rocks which resulted from the action of hot solutions forced upward and into the intrusive granites.

Any of the foundation rocks may be quartz. Because quartz can be fractionated, the hot gases and solutions that formed gold were forced upward, into and through it. Although quartz does not always bear gold, the miners soon learned that where there was quartz, there might well be gold. Thus, quartz was sometimes called the mother of gold. Large or small or irregular or straight, these veins of quartz were probably the source of gold-bearing gravel in the old rivers as well as the present-day rivers.

It was that continuous gold quartz belt of the Sierra Nevada that became known as the Mother Lode. To the north of the American River, the seams and veins split up and thus cannot be described as one mineralized belt or lode.

However, this is the story of the Sierra Nevada and the Mother Lode that can be read in the accounts left us by the gravels—the actions of the Neocene rivers, and the more recent rivers, digging away at the mountainsides and dropping their golden pebbles on the bedrock at the bottom of their channels. Stored away beneath the trees and foliage and lava flows, this gold from the ages was waiting to be discovered and brought out to make California the "land of gold."

When he was in his nineties Charley Peters wrote his autobiography as a life-long miner in the Mother Lode.

A forty-niner in his old age. Although most gold seekers were young men, some grew old before their time.

A gold miner's donkey, packed with all the supplies the miner would need for months, while working his claim.

II

El Dorado County

El Dorado County was one of the first twenty-seven counties the California Legislature created in February 1850. El Dorado County includes Coloma, where the first historic discovery of gold was made and which served temporarily as the county seat.

The name El Dorado means "the gilded one." It refers to the ancient story of an Indian chief in Columbia, South America, who was anointed and covered with gold dust during religious rites and who later washed off the gold in nearby Lake Guatavita. What was to become known as California's Mother Lode was one of the fabulous lands the Spanish Conquistadores had been searching for as they explored the New World, and the name El Dorado has been associated with any area producing gold. Various writers applied it to the California gold country during the gold rush. Bayard Taylor gave the name El Dorado to his famous book about his travels to California's gold region. A mining camp in the county and streets in many cities also bear the name El Dorado.

As news of the gold discovery spread, the miners rushed in and established diggings and camps along the American and Cosumnes Rivers, which became the northern and southern boundaries, respectively, of El Dorado County.

Many mining camps sprang up in the area as new discoveries were made. These camps lasted only a short time and then became ghost towns, and now are sources of interest to tourists and historians. Placerville and Georgetown are the only towns that remained sizable after the gold rush. Coloma was practically a ghost town until it became a state park in 1927. Since Coloma is where gold was first discovered and where the gold rush started, its history is vital to El Dorado and the entire Mother Lode.

The story of El Dorado County and the gold rush starts with the efforts of two men, Captain John Sutter and James W. Marshall, who were the key figures in the discovery of gold.

John Sutter, a Swiss-German, had had an unsuccessful dry goods business in Europe. He came to America in 1834 and eventually moved west to St. Louis. Here he heard stories of the great land of California and vowed to migrate there and obtain a big land grant from the Mexican Government. He joined an overland party to Oregon and from there sailed to California by an indirect route via Honolulu, Hawaii and Sitka, Alaska, arriving in San Francisco in 1839.

Sutter became a Mexican citizen, and from the Mexican Governor Juan Bautista Alvarado he received one of the largest Mexican land grants, eleven square leagues of land (or 48,000 acres) at the junction of the Sacramento and American Rivers. On this enormous rancho, which he called Nueva Helvetia ("New Switzerland") after his native country, Sutter was given complete power to enforce the law.

He planted orchards, vineyards and grains and recruited Indians for labor. In the center of the grant was his fort, with walls fifteen feet high and three feet thick. In 1841, for $30,000, Sutter bought all the properties of the Russians at Fort Ross and Bodega Bay, which he moved to Sutter's Fort.

To develop his community, Sutter needed a sawmill and timber, and to obtain these he turned to a carpenter or millwright named James Marshall, who had arrived from Oregon in 1845 and in 1847 was working for Sutter. He became Sutter's partner in building a small sawmill on the South Fork of the American River, about forty miles up the river, at a place the Indians called Coloma (Culloma). The sawmill was to be operated by waterpower and, in order to deepen the tailrace, Marshall had his crew loosen gravel on the bottom and run a swift current of water through it overnight.

While inspecting the tailrace the next morning, Marshall saw a nugget or flake of gold in the gravel. In a few minutes he had picked up several more flakes, and after testing them he called his men and showed them his discovery. He pledged the men to secrecy and headed for Sutter's Fort.

Among the mill crew were several men from the Mormon Battalion which had come overland in 1846. Before rejoining the Mormons at Salt Lake City after their release from service, several of them were working for Sutter. One of these Mormons was Henry Bigler, who wrote in his diary, "The boss of the mill found some mettle he thinks is goald." It was dated January 24, 1848. This notation established the gold discovery date for posterity, for James Marshall later thought the day was January 19.

That the shiny metal was in fact gold was confirmed by three tests: for weight, for malleability, and for its reaction to nitric acid, which has no effect on gold except to make it shine more brightly. After testing the nuggets, Sutter and Marshall returned to the mill. They asked the men to finish the mill and keep the gold discovery secret. However, word leaked out, and the news soon brought hundreds of gold seekers to Coloma and the American River. About two thousand miners were washing the gravel of the river by the summer of 1848, and with the rush of 1849, the population reached over ten thousand.

Their roles in this history-making event proved personally disastrous for both Sutter and Marshall. Sutter's workmen deserted him to dig for gold, and his rancho was taken over by hordes of squatters. By 1852 he was bankrupt and moved to his Hock Ranch on the Feather River, but the American courts declared this grant invalid. The California Legislature granted him a pension from 1872 to 1878 for his service in settling New Helvetia (later Sacramento) and to compensate him for his losses.

After he had petitioned the United States Congress for additional compensation, he moved to a Moravian colony at Lititz, Pennsylvania. There he died, in comparative poverty, in 1880.

James Marshall also suffered. He and Sutter had tried to acquire ownership of the land and timber at Coloma from the American government. They had sent a man to Monterey to try to obtain legal title from the military government, but this effort failed. Marshall tried unsuccessfully to charge a commission for any gold the miners found. He became embittered, feeling that the state owed him a great debt for his discovery.

In 1872 the Legislature appropriated a $200 a month pension for Marshall. He then moved to nearby Kelsey, built a blacksmith shop, and lived there the rest of his life. His last years were sad. The monthly pension was cut, then eliminated entirely in 1878. Marshall eked out a living at various menial jobs and by selling George Parson's book *The Life and Adventures of James Marshall, 1870*, a small picture of himself and his autograph card, all of which could be purchased for $1.35. An eccentric recluse, he died in 1885, at Kelsey.

However, he was not to be forgotten. In 1890 the Native Sons of the Golden West erected a monument to him on the hill above Coloma, with a heroic statue pointing to the spot in the river where the first gold nugget had been found.

Johann August Suter, or John Sutter (1803-1880), reared in Switzerland, was really the key figure in the discovery leading to the gold rush. He arrived in California in 1839 and established a colony on the present site of Sacramento.

The main building of Sutter's Fort in ruins before the restoration in 1891-1893. This is the only original structure in the present fort complex. The walls and other features have all been completely reconstructed.

John Sutter built a fort on his land grant and armed it with guns bought from the Russians when they abandoned Ft. Ross on the coast. Sutter's Fort was the objective of most of the early American settlers coming overland to California.

Sutter's Mill, 1853. By this time the mill was no longer functioning and was being used as a primitive dwelling. It is claimed that James Marshall is standing in front.

An exact replica of Sutter's Mill as built in the 1960s by the State Department of Recreation. It is accurately reconstructed and can saw boards while visitors watch.

Sutter's Mill, 1851. Coloma, California as shown in Charles Nahl's famous painting.

James Wilson Marshall, a millwright for John A. Sutter, discovered gold in the tailrace of Sutter's Sawmill on January 24, 1848, on the south fork of the American River.

A replica of Marshall's hillside cabin at Coloma, where he lived for years.

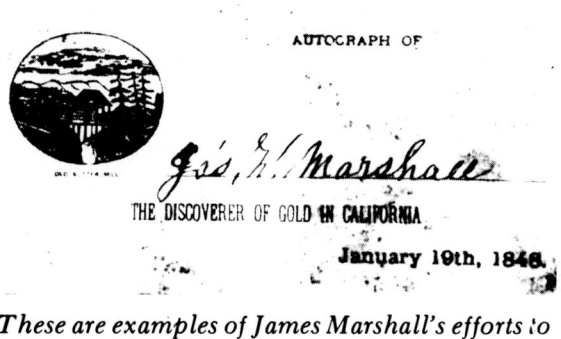

These are examples of James Marshall's efforts to gain recognition as the discoverer of gold in California. Notice the incorrect date on his autograph card.

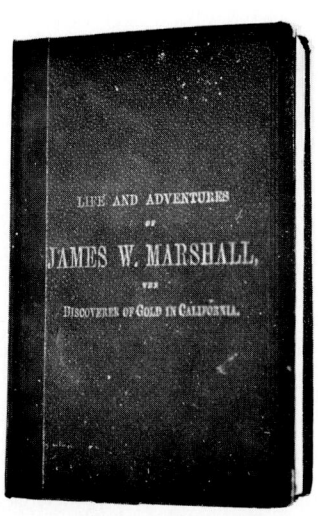

The Life and Adventures of James W. Marshall, the Discoverer of Gold in California
By George Frederic Parsons
Published in 1870
This little book, ghost written for James Marshall in 1870, was sold by him along with his picture and autograph, all for $1.35. Separately the book was $1.00, the picture 25¢, and the autograph 10¢. The book is now quite rare.

Marshall was buried on a hill overlooking the American River on August 12, 1885. This monument was erected in 1890, as a result of efforts of the Native Sons of the Golden West.

The San Francisco Foundry showing the men who cast the bronze statue of Marshall for the apex of the monument in Coloma. The figure of Marshall is seven feet tall.

Henry Bigler and several men from the Mormon Battalion who had come overland in 1846 to help in the conquest of California were working for Sutter and Marshall at the sawmill. Bigler kept a diary that has provided accurate data on the great discovery. He recorded the day as January 24, 1848.

Since the miners quite naturally headed for the locale of the original discovery, Coloma became California's first boom town. By 1852 the rush to Coloma was over and its most glorious period drew to a close. Thousands of gold seekers by then were exploring all the streams of the Mother Lode, from El Dorado County to Mariposa County. As the restless miners left for other diggings, Coloma became nearly deserted, and in 1857 the county seat was moved to Placerville, which had become a larger mining and commercial center.

Today, Marshall Gold Discovery State Historical Park, 220 acres in size, preserves about seventy percent of the original townsite of Coloma, including several old stone buildings and the cabin where Marshall lived. The state has built an excellent interpretive museum as well as an accurately reconstructed sawmill which park rangers operate for visitors. Because of concern for flood damage from the river, the sawmill runs by electric power rather than water power.

The state received title to the Marshall monument site in 1889 and the right to operate the area as a state park in 1927. Other areas and sites were added, and finally, in 1941, the gold discovery site was acquired. In 1948 a tremendous Centennial celebration was held here, and the first discovery of gold was re-enacted. This park may well be the most popular historical site in the California state park system, and the tourists who visit it have become the area's modern bonanza.

Placerville, the largest town in El Dorado County, had grown rapidly since 1848, when gold was discovered in the rivers there. Since water to wash the gravel was scarce, Placerville was called "Dry Diggings," but the gravel there was rich enough to attract large numbers of miners. As the place where several criminals were hanged, the diggings were often referred to as "Hangtown" until 1854, when the name Placerville was officially adopted and the town became incorporated.

After the rich placer mines were depleted, Placerville became an important commercial center and a stop between mining districts and points on the Overland Route. The Pony Express stopped here too. With the discovery and development of the Comstock silver strike at Virginia City, Nevada, in 1859, Placerville benefitted from the traffic, freighting, and reverse migration from California to Nevada. The Placerville-Carson Valley Road became heavily traveled, and Placerville grew in importance.

Today that wagon road over the Sierra is Highway 50, and Placerville is a tourist and recreational mecca, an industrial and agricultural center, and one of the largest and most active towns in the Mother Lode.

It is claimed that the Gold Bug mine in Placerville's Bedford Park is the only municipally owned gold mine in the world. Two tunnels are open to visitors, and nearby is a restored stamp mill.

Southeast of Placerville is Big Cut, one of the area's richest gold deposits. Giant hydraulic nozzles or monitors have cut down the cliffs here. It is claimed that one acre on the hill produced a million dollars in gold. By 1900, however, gold mining in this area had come to an end.

Georgetown, located northeast of Coloma, is another important community. Many authorities claim it is the northern end of the Mother Lode.

The name Georgetown apparently is taken from the George Phipps party that arrived here in 1848 and found the gravel rich with nuggets that "growled" in the miners' pans. (The miners claimed that the nuggets were so heavy that they made a "growling sound" while being washed out of the gravel. Because of this the camp was first called Growlersburg.) By 1851 the population had increased to six hundred men and six families, enough for a post office, which was given the name Georgetown. The town of tents and shacks was destroyed by fire in 1852. As it was rebuilt, wide streets were laid out, and stone and brick buildings were constructed.

By 1855 Georgetown had three thousand people and the cultural advantages of a town hall and theater. Today it is a recreation and commercial center. The wide streets, planned after the first big fire, are unusual in the Mother Lode, and a number of old buildings add to the town's appeal. One of these is the Armory, built in 1862 by the "Home Guard." It has survived over the years even though it has been adapted for various businesses. Other outstanding buildings are the Balzar House, a three-story hotel built in 1861; and three buildings from the 1850s: the IOOF Hall, the Opera House and the Shannon Knox House. Some writers call Georgetown "The Pride of the Mountains."

Near Georgetown are the remains of several early gold camps, including Kelsey, where James Marshall died in 1885. The town was named for Benjamin and Samuel Kelsey, who discovered gold there in 1848. Georgia Slide, Oregon Canyon, Cool Garden Valley and Pilot Hill are other nearby camps.

One of the few remaining nineteenth century structures, the Coloma Catholic Church was built in 1858 and is still in use after restoration.

The El Dorado County Museum, 100 Placerville Drive, Placerville, California. The museum is located on the El Dorado County Fairgrounds. The building was constructed in 1974 and is operated by the County Museum Commission. It is staffed by the Historical Society and is open Wednesday through Sunday from 10 a.m. to 4 p.m.

Main Street in Coloma in 1859. This picture was used as a model for the present Sierra Nevada House which was built by John Hasserler about 1959.

The sign says Sportsmans Hall was a stage station on the old Placerville road, twelve miles east of Placerville. The building was erected in 1853 by D. C. Deady and later purchased by the Blair brothers. This was a relay station for the Pony Express; it burned in 1868 and a replacement occupies the site. Some of the finest sugar pine lumber in the world come from this area. This site is State Historical landmark #704 on the Pony Express Trail at Cedar Grove.

The Combellack House at 3059 Cedar Ravine Street. The house was built by William Combellack in 1895 and is one of the most outstanding old homes in Placerville.

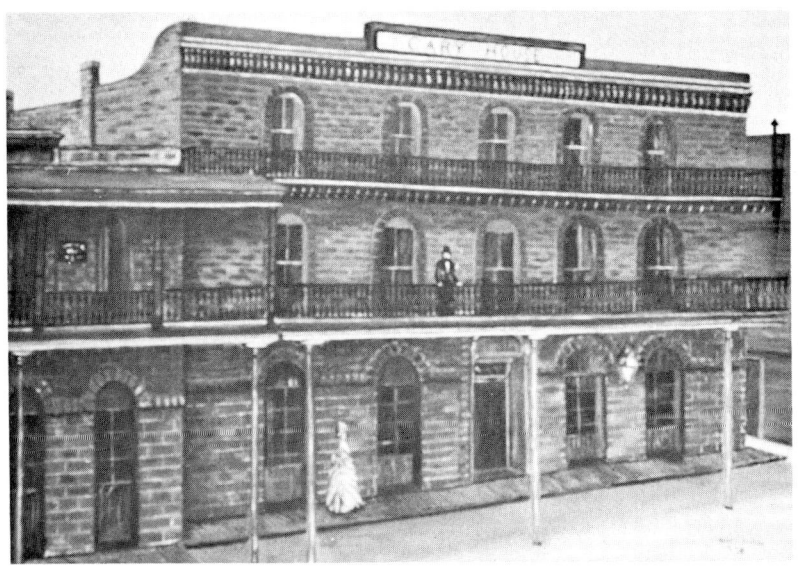

An artist's rendering of the reconstructed Cary House in Placerville, 1979. The original was built by William Cary in the revival of the town following a fire in 1857 which practically wiped out the town. It was used as headquarters for the stage lines as well as for Wells Fargo and Company, and it is said that during those years $90,000,000 in bullion from Nevada rested on its front porch. It was from the balcony of the Cary House that Horace Greeley delivered his presidential campaign speech to the miners and businessmen in the street. Before being torn down in 1911 the hotel changed owners eighteen times. It is being rebuilt on the original site.

An 1890s view of the Wells Fargo Express Office in Placerville. It housed the Alta Telegraph *and was located on the north side of Main Street about where the post office is located today.*

Parade on the main street of Placerville, 1871.

Overall view of Placerville about 1915.

John M. Studebaker (above) driving an electric car in 1903. He was a Placerville pioneer blacksmith, soldier, inventor and builder. The Studebaker building (at left) stood on Main Street in Placerville from 1853 to 1858.

Placerville and the surrounding area produced about $70,000,000 in gold up to about 1880. Originally known as Hangtown, the town was incorporated in 1854 and became the county seat in 1857. Among the well known men who contributed to Placerville's early history were: John M. Studebaker, who built wheelbarrows in Hangtown and eventually became a giant in the automobile industry; Snowshoe Thompson, who carried mail on skis over ninety miles between Placerville and Carson City, Nevada for more than a decade; Mark Hopkins, who established a store in Placerville with a wagonload of supplies and later became one of the Big Four of railroad fame; Hank Monk, a famous stage driver who drove Horace Greeley over the Sierra to Placerville; Sheriff James Hume of El Dorado County, who served as a detective for Wells Fargo for twenty-eight years; Philip Armour, operator of a butcher shop and later of the famous Chicago packing company; Captain Charles Weber, the founder of the city of Stockton, who successfully operated the Stockton Mining Company on Webber [sic] Creek near Placerville in 1848; and Horace Greeley, the editor and politician, who made a campaign speech from the balcony of the old Cary House to the miners of Hangtown when he was his party's presidential candidate.

Hank Monk, known as the "Whip," was a famous stage driver. He brought Horace Greeley to Placerville in 1859 while Greeley was on a campaign tour of California.

Nine-foot long skis used by John A. "Snowshoe" Thompson, who started carrying mail across the Sierra between Placerville and Carson City, Nevada in the winter of 1856-57. A man of tremendous endurance, he made the trip to Carson City in three days and back to Placerville in two days, and continued the routine into the 1870s. His normal load was from sixty to eighty pounds. He was never paid by the Federal Government.

A view of Georgetown in the 1920s.

Shannon Knox House, constructed in the early 1850s, was saved by hard fighting in the great fire of 1852, which destroyed the rest of the town. It is located on the southwest corner of El Dorado and Main Streets in Georgetown.

The present Georgetown Hotel was built in 1896 and carries on the pioneer tradition of friendliness and good food. The original hotel was built in 1852 and burned down only four years later.

Georgetown was first prospected in the summer of 1849. After the fire of 1852 the town was rebuilt with streets sixty feet wide to prevent total destruction in the event of another fire. Fires did occur again, in 1858, 1869, 1873 and 1897. By 1855 the town boasted over seventy businesses, and had telegraph and stage connections to Sacramento, Placerville, Auburn and towns farther north towards the Middle Fork of the American River.

The St. James Catholic Church in Georgetown is still in use today, although a new church is to be built in 1980.

The Balzar House, originally a three-story brick building, was built in 1861 for a Mrs. Olmstead, whom the miners called Widow Balzar. She operated the building as a hotel with a dance hall on the top floor. The venture failed, however, and she sold out to Joseph Whiteside in 1870. Whiteside, at great expense, turned the building into an opera house. In the late 1880s it was purchased by the Odd Fellows, who dedicated it as a meeting hall in April of 1889; the following year they removed the top floor. Today the building stands vacant on the northeast corner of El Dorado and Main Streets in Georgetown.

A drawing of the Wakamatsu Castle of Aizu, Japan decorates the wall of the Gold Hill Trail Elementary School. The Wakamatsu Tea and Silk Colony at Gold Hill brought the first Japanese to California in June 1869. They had plans for large scale cultivation of tea and silk, and brought thousands of mulberry trees, bamboo roots, tea seeds, and grape cuttings. Farming and house building were begun when other colonists followed and for a while the project had a promising outlook and created a stir among county residents. Despite the initial success it failed to prosper; nevertheless, it marked the beginning of Japanese influence on agriculture in California.

The only evidence left of this pioneer Japanese settlement at Gold Hill is the grave of "Okei," a nineteen-year-old Japanese girl who died of typhoid fever in 1871. The gravesite overlooks the Gold Hill Trail Elementary School. A monument in memory of Okei is on Mount Seaburi, Japan, a gift from the mayor of Aizu Wakamatsu, Japan.

Wakamatsu Tea and Silk Colony at Gold Hill. Beautiful Japanese landscaping surrounds the historical marker near the Gold Hill School yard. The plaque was presented by the Department of Parks and Recreation in cooperation with Japanese American Citizens League of El Dorado County, the Historical Society and Friends. It was dedicated on the Centennial of the first Japanese settlement in California.

Quartz Mining

The first gold-seekers rushed to Coloma and the American River, where they found free or placer gold in the streams. Very few of them had ever done any mining, and they had little equipment for the work except hand tools and a pan or a home-made gold rocker. Most of the early gold camps in what became El Dorado County were based on placer mining, but some miners were looking for the sources of the free gold in veins or lodes. They found outcroppings of the Mother Lode quartz veins in several places but discovered that these terminated in the vicinity of Georgetown. After meandering northward about 165 miles from Mariposa County, the gold quartz veins and the Mother Lode become shattered, near Georgetown, but the Sierra Gold Belt continues northward through Placer, Nevada and Sierra Counties some 275 miles, although it cannot be traced as a continuous lode.

Just north of the Cosumnes River, in El Dorado County, was Nashville, claimed to be the first hard rock mining camp in the state, and the Tennessee-Nashville mine, where the first stamp mill in California is said to have been operated. Mariposa also claims the state's first stamp mill, built in July 1849 by Kit Carson and two associates, who discovered the Mariposa Quartz gold mine in the spring of that year. Stamp mills were put into use when miners found that the Mexican *arrastras* were not adequate to crush the quartz gold ore.

It would be safe to say that the Tennessee-Nashville was the first hard rock mine in El Dorado County and one of the first in the state. Available records show that this mine produced over $230,000 in gold, which is not unusually small as the quartz mines of El Dorado did not produce as much gold as those in the Southern Mines along the Mother Lode.

Many other hard rock mines were developed just to the north of the Cosumnes River. One was the Montezuma, and then to the northeast of it were several mines such as the Red Bird, Union, and Martinez. The Union is credited with being the biggest producer in El Dorado County, with a yield of over $2,700,000.

Nearby, on Highway 49, is the location of the mining camp of El Dorado, originally called "Mud Springs" because its water was constantly muddied by livestock. It became an important crossroad and stop on the Carson Emigrant Trail, and in 1855 became the incorporated city of El Dorado. After the placer deposits were exhausted, El Dorado continued to thrive on the lode mines and quartz mills lining the roads to the Cosumnes River. Nearby Diamond Springs was a placer mining camp which had no important quartz mines.

Several lode mines were developed near Placerville. The Pacific quartz mine, located on a hill south of town, was the richest. The records show that it produced $1,486,000 in gold. Coloma had no important quartz mines, and in nearby Lotus such activity played out after a brief period. On the road to Georgetown is Garden Valley, which had the Black Oak Mine, one of the best in El Dorado County, with a production of over one million dollars in gold.

At Georgetown the two best-known hard rock mines were the Alpine, in the center of town, and the Beebe, at the northern edge. Both produced occasional rich ore streaks.

No one has any reliable figures for the total gold production of El Dorado County, as its fame was based largely on placer mining and individual efforts, but this is where gold was discovered and where the hordes of gold seekers headed at the beginning of the gold rush. A rich heritage of those exciting days is evident now, and large numbers of tourists visit each year. On special occasions, such as Gold Discovery Day, ten thousand visitors may come to Coloma, emphasizing the importance of the past and our fascination with it.

The headframe and tailings dump of the Nashville Mine, located six miles south of El Dorado on Highway 49. In his book, California Pioneer Towns, *Art Lassagne states that Nashville was one of the first quartz mining areas in California. The booklet "Historic Sites of El Dorado County" states that Nashville was the first real hard rock mining camp in El Dorado County and was first known as Quartzville.*

The Georgia Slide mine has been worked and has paid dividends intermittently since 1853. It is the oldest continually worked mine in El Dorado County and probably in the state. It rises 300 feet above Canyon Creek and has a working face 200 feet in height. Said to have been named for Georgian miners in 1849, it is located on the south bank of Canyon Creek, north of Georgetown and east of the road to Bottle Hill.

This picture of Grizzly Flat is from an old stereograph. Located about twenty-five miles east of Placerville, Grizzly Flat was one of the richest mining districts in El Dorado County, with hydraulic, placer and hard rock mines. Beginning about 1850 hotels, blacksmith shops, churches, various saloons, stores, and private homes were established. It was given its name after a prospector killed a large grizzly bear that raided his camp. The town burned twice, once in 1866 and again in 1869. After the last time it was never completely rebuilt. Within a few miles of the town were sawmills, ore-crushing mills and well established mines. An attractive modern subdivision has been constructed near the townsite.

LUMBER INDUSTRY
In El Dorado County

Although El Dorado County is the place where gold was discovered, its deposits proved less rich than those of other Mother Lode counties. A much more valuable resource, in fact, has been timber, which by now has exceeded many times the value of gold produced in the mines. All of the Mother Lode counties located on the foothills of the Sierra Nevada Mountains have large stands of pine forests, but El Dorado County has a particularly rich endowment.

In a way the timber industry was actually responsible for the discovery of gold, for Captain Sutter had sent James Marshall to build a sawmill at Coloma, on the American River, to cut the fine forest of pine located there. This timber would produce the lumber Sutter needed at New Helvetia, or Sutter's Fort. That sawmill was completed, even though gold was discovered in the process.

This was the beginning of the great timber industry in the county, which now produces various types of products ranging from heavy building timbers to the output of numerous box factories. It is a multi-million dollar industry employing several thousand people.

Over sixty-five percent of El Dorado County is timbered, with some areas containing the finest stands of sugar and Ponderosa pine, white and Douglas fir, and incense cedar to be found on the Pacific slope.

The annual cut in El Dorado County exceeds 150,000,000 board feet and supplies over fourteen sawmills. Controlled cutting in the El Dorado National Forest assures that there will be a perpetual supply of good lumber because only the mature trees are logged, so by the time the area is cut over, a new growth will be ready for harvesting.

Lumbering on private lands is important too. The popular Western Pine Association Tree Farms, begun in El Dorado County in 1944, are doing a great deal to assure perpetual timber crops. Today as many as 147,500 acres in El Dorado County are under this form of management.

The Institute of Forest Genetics of the United States Forest Service is located on a 106-acre site four miles east of Placerville. Here laboratories experiment with producing hybrid trees which not only grow more rapidly but also are resistant to insects, disease and drought. Nowhere in the world is there a larger stand of pines for experimental purposes than the one here.

In 1958 the Placerville Tree Nursery of the El Dorado National Forest was established to provide the eighteen national forests in California with seedlings of Douglas Fir and Ponderosa, Jeffrey and sugar pine. Three million seedlings were produced the first year, but production has tripled since then, and now more than fifty people are employed full-time during the tree lifting and packing seasons.

Forest Service personnel collect the seed from the various national forests. Several interesting and valuable pine trees have been developed, and these will markedly increase the value of timber to the economy of El Dorado County.

Sawmill tools used in Pino Grande are on display in the El Dorado Museum in Placerville.

An ox-drawn log wagon arrives at Blair's Sly Park Sawmill in 1890.

The picture at the top was taken at Pino Grande, which is north of the South Fork of the American River. The sawmill was then called the El Dorado Lumber Company. In 1911 it became the C. D. Danaher Pine Company and then, in 1918, the Michigan-California Lumber Co.

In the 1940s the sawmill was dismantled and moved to Camino, about twenty-four miles northeast of Placerville. A recent photo of the sawmill is above.

The sawmill at the C. D. Danaher Pine Company (now the Michigan-California Lumber Co.).

Workmen unloading logs into a pond at the C. D. Danaher Pine Company, circa 1915.

Hauling the products of the C. D. Danaher Pine Company Mill.

A Diamond and Caldor Railway logging train. There were many trestles in El Dorado County, but this one's 97-foot length made it one of the longest ever erected for a lumber road up to that time. The bridge had a total of sixty-three individual supports, with a combined length of 10,992 feet.

Automobiling on rails, Diamond and Caldor Railway.

A Diamond and Caldor locomotive displayed at the El Dorado History Museum in Placerville.

The donkey engine at right is snaking logs down the "skid road" to a railroad loading area. This picture shows one of the Caldor operations.

A donkey engine at work in one of the Caldor logging camps. The horse would pull the cable into the woods. There the cable was attached to logs which the donkey engine pulled to the railway. (The donkey engine was mounted on a sled-like platform so that it could pull itself from location to location as needed.)

Hauling logs in the winter in the Sierra, circa 1930s, using a sledge and a Caterpillar Model 30 tractor.

Hauling Ponderosa pine logs to the Thatcher Mill near Shingletown in 1901.

Log pond at Pino Grande.

American River Land and Lumber Company drovers recovering logs from shallow water and rolling them into deeper water, circa 1889. This location is on the American River in El Dorado County, on the way to Folsom.

Unloading operations on the Diamond and Caldor Railway about 1900. These logs were sawed into lumber for a large door factory in Oakland, California. The Caldor Company had 30,000 acres of timber in El Dorado County.

III

Amador Area

Amador County was not among the first twenty-seven counties created by the California Legislature in 1850, for this gold-laden area along the Mother Lode was originally part of Calaveras County. It was not until 1854 that Jackson and the area north to the Cosumnes River developed sufficiently to break away from Calaveras County and be recognized as a separate county, with Jackson as the seat of government. Later this county was to annex territory from El Dorado County, and in 1864 it lost a large part of its eastern territory with the formation of Alpine County.

Jackson was originally known as Botellas (bottle) Spring because early "litterbugs" had scattered dozens of bottles around the future townsite. A spring located here was a good watering place for miners and teamsters, and the town grew around it. It was later named for Colonel Alden Jackson, who mined there for a short period and was an effective, energetic leader during the settlement's earliest development.

Jackson was not a rich area during the early placer mining years (1848 into the 1850s), but it was located in an opportune place to become a commercial center for richer mining areas nearby, including Big Bar, Middle Bar, Oregon Bar, and others along the Mokelumne River.

When the quartz veins of the Mother Lode were discovered in the 1850s, it was found that the main lode ran through the center of Amador County and directly under Jackson, giving rise to the claim that "more than half the gold mined in the entire Mother Lode came from the mines of Amador County." If that is so, then Amador County is truly "the heart of the Mother Lode." Certainly some of the richest mines on the Mother Lode were discovered and developed in this area. These included the Argonaut and the Kennedy at Jackson; the Central Eureka, Union and Lincoln in Sutter Creek; the Consolidated Keystone mine in Amador City; and the Plymouth Consolidated at Plymouth.

Jackson's chief competitor during the early years of the gold rush was Volcano, discovered by the discharged soldiers of

Colonel Jonathan Stevenson's New York Volunteers. The total value of gold produced there has been estimated at $90,000,000. This figure also includes the later hydraulic period since there were no deep quartz mines in or around Volcano.

Sutter Creek, once famous for the Central Eureka Mining Company, was named for Captain John Sutter because in the 1840s he had cut timber and also mined there. Sutter Creek has one of the best preserved main streets in the gold country, with many buildings still retaining their old wooden balconies, facades and gingerbread.

Gold mining here and throughout California ended in 1942, when the United States Government closed down such operations. Later the government repealed its order, but few mines reopened because of the low price of gold. One that did was the Central Eureka, which operated until 1968 before economic problems forced it to close.

The Knight Foundry was established by Samuel N. Knight in 1873 to serve the mines of Sutter Creek and is still in operation. The foundry specialized in mining machinery. One of their products was a widely accepted water wheel that was used in running ore mills and a variety of other operations, including the foundry itself. Even today, if need be, the Knight Foundry could be operated by these cast iron water wheels.

Amador City, north of Sutter Creek, is also a picturesque community. It was made famous by the Keystone Consolidated mine, an excellent producer during a remarkably long active era extending from 1851 to 1942. It produced $24,000,000 in gold before it was forced to close.

The Empire Mine yielded $7,000,000 in a period of five years and became a part of the Plymouth Consolidated Mining Company, helping to make Plymouth an important mining center on the Mother Lode. Today it is a commercial center and the site of the beautiful thirty five-acre Amador County Fairgrounds.

At Plymouth the traveler along Highway 49 can turn east to the Shenandoah Valley. Vineyards there produce some of the finest wines in California. The D'Agostini family operates the oldest winery in the valley, and their wines are famous throughout the West.

Nearby is the quaint little mining town of Fiddletown. In the 1870s its name was changed to Oleta because of political pressure. Many years later the colorful original name was restored, much to the delight of tourists and residents alike.

The gold mining industry has given Amador County a rich and romantic history, tangible evidence of which is available to visitors today. Many people come each year to visit such interesting and historic sights as the famous Kennedy mine tailing wheels.

As tourists learn the story of the wheels, they are given the history of the Argonaut and Kennedy mines whose head frames are still standing and visible from the highway. These mines produced over $59,000,000 in gold and were the deepest in the country at one time. They also have a dramatic history. Forty-seven miners lost their lives in the Argonaut in a 1922 cave-in, despite efforts to rescue them via the Kennedy shafts and tunnels.

The visitors who travel Highway 49 to these little mining towns are very important to Amador County's economy and may bring in more wealth than the yellow gold from the mines did originally.

Scenic Highway 88 over Carson Pass also brings people to Amador County. Kirkwood, one of the best ski centers in the West, is on this highway, 7,800 feet high. At a lower altitude, beautiful Daffodil Hill, near Volcano, is a colorful attraction, especially in the spring. Nearby Indian Grinding Rock State Park is rapidly becoming a major California Indian cultural center and tourist attraction. Near the valley, the Camanche and Pardee reservoirs along the Mokelumne River offer fishing and water sports. Amador County truly offers its visitors a variety of pleasures!

The lumber industry is another vital component of Amador County's economy. The big sawmill at Martell, near Jackson, is operated by the American Forest Products Company, a subsidiary of the Bendix Corporation. It was built in 1941, but since then the operation has been expanded to include the manufacture of plywood, particle board and moulding. There is also a bark processing plant and waste wood chipping facility to produce chips for paper manufacturing. To service this big plant the American Forest Products Company brought the twelve-mile line of the Amador Central Railroad to connect with the Southern Pacific at Ione.

The company harvests lumber from 126,000 acres of forests that are scattered throughout Amador, Calaveras and El Dorado Counties and also from nearby National Forest Lands. The purchase of timber from the U.S. Forest Service amounts to about $3,000,000 annually.

The yellow gold of the mines developed the Amador area, but the gold that supports it today is the tourist gold brought by those who visit the historic and recreational sites, the green gold of the lumbering industry, and the agricultural gold from the farms. These resources are as valuable as the yellow metal in the mines and give the area a more broadly based economy.

Colonel Alden Moore Jackson, after whom the town of Jackson was named.

The original Jackson elementary school was built in 1860. In the 1950s it was replaced by a new building, located across the street from the Amador County Museum.

An early view of the city of Jackson, with Jackson Butte in the background.

The city of Jackson in 1979. Only Jackson Butte looks the same!

The Amador County Courthouse in Jackson, built in the 1860s, has long since been replaced.

The stage in front of the Louisiana House (now the National Hotel), at the south end of Main Street in Jackson.

A 1920s view of the National Hotel, located at the end of Main Street in Jackson. Originally called the Louisiana House, it has been in operation since 1863.

St. Sava's Serbian Orthodox Church is located on Jackson Gate Road. Erected in 1894, it is one of Jackson's most remarkable buildings. Jackson has a large Serbian population, and St. Sava's is the Serbian Mother Church in the Western Hemisphere.

St. Patrick's Catholic Church in Jackson has been in continuous use since 1868.

The Brown House in Jackson was built in 1859 by Armsted C. Brown, an attorney and first president of the Jackson city trustees. Brown served as a state Assemblyman throughout much of the 1860s and as a county judge from 1876 to 1879. The Amador County Museum acquired this stately house as its home in 1949. In 1920 Will Rogers filmed "Boys Will Be Boys" at this house.

On September 11, 1886, a group of young women met in Jackson and formed the Native Daughters of the Golden West, which became a statewide organization. This photo of some of the charter members was taken in April 1933.

A 1920 view of the old Pioneer Hall on Main Street in Jackson, where the Native Daughters of the Golden West was organized in 1886.

Lily Reichling, founder of Jackson's Ursula Parlor of the Native Daughters of the Golden West, in a 1932 portrait.

Originally called the Pine Grove House, the Pine Grove Hotel was built by Albert Leonard in 1855. The hotel is now long gone, but this view was taken about 1916.

The main street of Pine Grove many years ago. The town was started in the early 1850s and was primarily a stage stop. Pine Grove is famous mainly as a temperance center; its anti-liquor leaders even tried (unsuccessfully) to take a woman to court for putting liquor in a mince pie!

Mac's Place in Pine Grove was a general store, post office, and stage stop on the Carson Pass Road, one of the popular emigrant roads over the Sierra to the gold country. Kit Carson guided John C. Fremont over this pass in 1844. Pine Grove is nine miles east of Jackson on Highway 88.

Drytown, four miles south of Plymouth on Highway 49, was established in 1848 and is the oldest town in Amador County. Its population is said to have reached 10,000 at one time. "Drytown" was a misnomer, for its twenty-six saloons were readily available to the thirsty throngs!

Volcano's Main Street in the 1920s and in 1979. Volcano is thirteen miles east of Jackson. Its diggings yielded over $90,000,000 in gold and were discovered in 1848 by the men of Colonel Jonathan Stevenson's New York Volunteers, who worked the area known as Soldier's Gulch. Volcano's name comes from the miners' belief that the townsite was located in the crater of a volcano. The town once had a population of over 8,000 and supported more than thirty-five saloons, two breweries, twelve restaurants and four hotels.

The three-story St. George Hotel, built in 1862, is a grand old Mother Lode hotel, with balconies running the length of the two upper floors. The hotel is still in operation today.

Part of the Masonic Cave in Volcano, where the first five meetings of the local lodge were held. This picture, taken many years later, includes some of the original members. This historical site probably is the only place in California where the Masons met in a cave.

This old brewery, built in 1856, at one time supplied the thirty-five saloons in Volcano. The building still stands today.

One of Volcano's most interesting attractions is "Old Abe," the cannon that helped win the Civil War without firing a single shot. When local Secessionists threatened to take over the town, the Volcano Blues, a home guard, acquired the cannon to confront them. They had no cannon balls but planned to use round rocks from the stream beds. This show of force restrained the Secessionists.

After the war the cannon became a source of great interest and was borrowed for an exhibit by some Sacramento people. The Sacramentans mistakenly thought it was a gift rather than a loan and so failed to return it. It took an all-out effort by the people of Volcano, through the Legislature, to get the cannon back.

In May 1931, a big celebration was held in honor of the cannon's return. San Francisco Mayor Angelo J. Rossi was there, and "Old Abe" was finally fired—with real cannon balls!

Placer gold was discovered in the summer of 1848 at a site to be named Sutter Creek, in honor of John Sutter. Many of the early residents were of Italian descent, with the Monteverdes, Malatestas, Le Vaggis, Brignoles, Belottis, Cavagnaros, and many others settling there. For years the annual Italian Picnic on the Fourth of July has been a major event in Sutter Creek.

Many rich mines developed around Sutter Creek, and some say one of the first stamp mills in the Mother Lode was there, at the Union Mine. Leland Stanford bought the Union, renamed it the Lincoln, and turned it into a rich producer. Stanford used the profits from this venture during the 1860s to invest in the Central Pacific Railroad.

The Sutter Creek Inn stands where John Keys first built a home for his bride, Clara McIntire, in 1860. The fine house on Spanish Street, at the foot of Haydon Alley, burned to the ground the day the Keyses were to move in. They then built a smaller house on the same site, where they lived for the rest of their lives. Their house is still there; today it is the Sutter Creek Inn.

In 1964 these three unidentified men re-opened the old Central Eureka Mine. It was the only working Mother Lode mine during the time it was in operation. The venture was not successful, however, and the mine has since closed down.

An early view of the Knight Foundry at Sutter Creek, established by Samuel N. Knight. The main purpose of this foundry was to manufacture Knight's water wheel (Samuel Knight's invention, rivaled only by the Pelton wheel). The foundry has never ceased operation since it was established in 1873 and continues to serve a large area.

The old Central Eureka Mine, located on Amador Road past the Union-Lincoln Mine. The Central Eureka was a combination of several mines, including the South Eureka and Old Eureka or Hetty Green. In 1899 the South Eureka became the first mine in Amador County to use electric power exclusively. The Central Eureka was converted from the use of wood to the use of crude oil in 1902.

An early picture of Amador City, looking north from the Keystone Consolidated Mine. The town was named for Jose Maria Amador, a soldier from the Presidio in San Francisco. Amador established a camp on this creek, with the help of friendly Indians nearby, and mined for gold in 1848. When the county was formed in 1854, it too was named Amador.

It is claimed that four ministers of the gospel were mining placer gold here and that one of them, Reverend Davidson, a Baptist, discovered quartz gold in 1851 in a gully that later became Minister's Gulch. This is said to be the original hard rock discovery in Amador County.

A 1979 view of Amador City, looking north along Highway 49. In the background on the far right is the Amador Hotel, built in 1856. At one time Amador City had a population of 10,000 gold seekers, but now, with a population of 202, it is the smallest incorporated city in California.

Amador City's Keystone Consolidated Mine and 60-stamp mill, looking east across String Bean Alley from the mine office. The mine shaft was 2,680 feet deep along an average incline of fifty-two degrees. Between 1851 and 1942 it was one of the most productive mines in the Mother Lode, and over $24,000,000 in gold was shipped before it closed.

Plymouth Consolidated Gold Mining Company produced $13,000,000 in gold during its operation. Its headframe and tailings are visible from Highway 49. The mill at right, built in 1913, was the first in the state to have a mine-to-mill conveyor.

A 1919 view of Plymouth, formerly known as Pokerville. Plymouth started as a gold-mining community but became a trade center for farms, ranches and vineyards in the nearby Shenandoah Valley.

Plymouth was noted for its rich Empire Mine (not to be confused with the Empire Mine of Grass Valley). The Plymouth Empire Mine, 1,800 feet deep, was connected to the Pacific Mine, which in one five-year period yielded $7,000,000 in gold. In 1883 the two mines were formed into the Plymouth Consolidated Mining Company. Operations here continued until 1947.

The D'Agostini Winery was started in 1856 by the Uhlinger family, who had migrated from Switzerland. The winery is about eight miles east of Plymouth. It has been in the D'Agostini family since 1911 and has been designated a California Historical Landmark.

A 1910 view of 20,000 non-irrigated grapevines adjoining the D'Agostini Winery. Some of the original vines are still producing.

These old oak casks are part of the present winery, which produces claret and sauterne and an estate-bottled zinfandel that is its finest product.

Four D'Agostini brothers—Armenio, Michele, Tulio and Henry—operate the winery today. They produce a limited amount of wine, Armenio says, and delivery is limited to a 150-mile radius.

Tools and hand pump used by Jacob Uhlinger who made his own casks when the D'Agostini Winery was founded.

In 1848 Missouri prospectors settled at the site of Fiddletown, which is located seven miles east of Plymouth off Highway 49. The miners spent so much time swapping yarns and fiddling that the town became known as Fiddletown. Bret Harte added to the community's fame with his story, "An Episode in Fiddletown." The photo on the left is of Fiddletown in the 1920s. From about 1870 to 1936 the town was known as Oleta.

The Kennedy Mine was 5,912 feet deep when it closed. It was the deepest mine in the United States. When the mine was active, mule-drawn ore cars were used in its 180 miles of underground workings. The Kennedy had a total production of about $35,000,000.

On September 7, 1928 the Kennedy Mine suffered a surface fire that destroyed the wooden headframe and all the plant buildings except the mill and main office. These were soon rebuilt and back in operation, but shortly after they were completed, another fire burned the headframe and adjoining buildings. As a result of these fires, a steel headframe with ore bins and primary crusher was built. This structure cost $210,000 and still stands today.

A look at one of the four old tailing wheels, circa 1950, with the headframe of the Kennedy Mine and Milling Company in the background. Starting in 1914 these wheels lifted the liquid tailings or waste from the mill over a ridge for impounding behind a dam. The tailings were elevated forty-eight feet. Each wheel was sixty-eight feet in diameter and had 176 buckets; 500 tons of solids could be handled. The wheels were driven by an electric motor and last ran on November 4, 1942.

At the north end of town is Jackson Gate, home of the two survivors of a group of four picturesque tailing wheels used to carry tailings from the 100-stamp mill at the Kennedy Mine. These two have been preserved in a county park that attracts many visitors.

The Argonaut Mine was very small when it was started in 1850, but it operated continuously from 1893 to 1942 with inclined shafts to a depth of 6,300 feet, or 5,570 feet vertically. The mine produced over $25,000,000 in gold before it closed in 1948. The headframe above the shaft survives still.

In the Argonaut cave-in disaster of 1922, forty-seven miners were trapped in the depths of the earth. Rescue crews tried to open a passage from the Kennedy Mine nearly a thousand feet away, but unfortunately they were too late. The trapped miners built three bulkheads to try to keep out poison gas, but their effort failed. This message on the wall reveals some of the agony of the entombed men. It was written with a smoking carbide lamp by one of the trapped miners, whose name remains unknown.

Carbide lamp and safety helmet used by miners in the 1930s and 1940s.

A working scale model of the Kennedy Mine Surface works is on display at the Amador County Museum in Jackson.

Daffodil Hill, three miles northeast of Volcano, is a blaze of color in late March and early April. For years the McLaughlin family has planted thousands of bulbs with over three hundred varieties, many from Holland, in the five-acre site.

The main bed rock mortar group in Chaw'se or Indian Grinding Rock State Park located on the Pine Grove-Volcano Road. This grinding rock measures 173 by 82 feet and is covered with 363 petroglyphs or rock writings, as well as 1,185 chaw'ses or mortar cups. They were used to pulverize acorns and other seeds that were a basic source of food to the Miwok Indians. A superbly reconstructed dance or round house can also be seen at the park.

A view of Ione in the 1920s. Founded in 1848, Ione served as a supply center first for the mines and then for the farms and ranches of the fertile Ione Valley. The town's original name of Bedbug was later changed to Freezeout, but it is claimed that local residents were so embarrassed by the name Freezeout that they changed it to Ione, after the heroine of Bulwer-Lytton's novel The Last Days of Pompeii.

Today Ione is the second largest city in Amador County, with a population exceeding 2,400. In recent years the facades of several buildings along the main street have been reconstructed, using ornaments salvaged from other buildings, to create a frontier look.

Martell, on Highway 49 west of Jackson, is the home of the Amador-Calaveras Division Plant of American Forest Products, a subsidiary of the Bendix Corporation. This is the largest sawmill in the Mother Lode. This enormous pile of logs was harvested from the 125,000 acres of forest the company owns in Amador, Calaveras and El Dorado counties, as well as nearby National Forest lands. The logs are sprayed constantly to keep them from drying out.

The Ione Methodist Episcopal Church was started in 1862 and completed in 1866. Dedicated as "Ione City Centenary Church" and later popularly known as the "Cathedral of the Mother Lode," the church was the first to serve the religious needs of the people in the area and is still in service.

Early pictures of the "castle" at the Preston School of Industry at Ione. The act creating the Preston School was passed in 1889. Under construction from 1890 to 1894, the "castle" originally contained about 120 rooms and had a swimming pool in the basement. The facility was an educational institution rather than a reformatory, and for a time was nearly self-supporting, with its many acres of agricultural land and manufacturing projects. The main building has been abandoned since 1960; facilities around it are now operated by the California Youth Authority.

The sand plant near Ione which supplies sand for several glass plants, including Owens-Illinois in Tracy. Owens-Illinois is a major manufacturer of products dependent on the unique Ione formation.

In Amador County is a unique source of gold, black gold taken from the deposits of lignite (a soft form of coal) that runs across the county from Ione to Buena Vista. From this lignite is extracted mantan wax, which is used in carbon paper, floor wax and polish. Most of this wax is imported from Germany, but during World War II, when the supply was cut off, this plant was developed. One place in Arkansas mines this type of lignite, but it is inferior to the Ione deposits. This now is the only active coal mine in California. In the latter part of the nineteenth century coal was mined extensively and in 1877 the Central Pacific Railroad Company built a branch line from Galt to haul out the coal.

Interpace Corporation of Ione mines and produces valuable refractory and ceramic raw materials from high-grade clay. The area has been a sand and clay producer for many years. The Dosch pit has been worked, it is claimed, since 1864, and is the oldest continuously used clay mine in California.

Miners taking a break in front of the entrance to a clay mine in Ione. Notice the hand auger that one of the men is holding.

IV

The Calaveras Region

The name Calaveras apparently goes back to Gabriel Moraga's 1806 exploration trip into the Central Valley in search of sites for new missions. When Moraga came upon a pile of skulls on a riverbank, he recorded the spot on his chart as "Rio de las Calaveras," or River of Skulls. The name survived, and when the first twenty-seven counties were organized by the California Legislature in 1850, the county took the name of the river. The name has also been applied in the diminutive form, meaning "little skulls," in Calaveritas Camp and Creek.

The present boundaries of Calaveras County are the Mokelumne River on the north and the Stanislaus River on the south. Originally the county was a good deal larger. Most of Amador County, in fact, was created from Calaveras County in 1854, and parts of Calaveras County were given to Alpine and Mono Counties when they were created.

Gold brought miners to the Calaveras area in large numbers in 1848 and 1849, to wash the sand and gravel bars of the Stanislaus, Calaveras and Mokelumne Rivers and their tributaries. Gold pans, rockers, long toms and sluice boxes were all that was needed to mine the abundant free gold, and this region soon became famous as the richest part of the Southern Mines.

The Mokelumne River gravel bars, especially, were unbelievably rich, and gold was easily washed out at such places as Oregon Bar, Middle Bar, Rich Bar, and Big Bar. The most important of these was Big Bar, on the Mokelumne River where Highway 49 crosses it today. Big Bar, where mining began in 1848, led to the discovery of rich gravel in the Mokelumne Hill area, the richest early placer mining camp in Calaveras County.

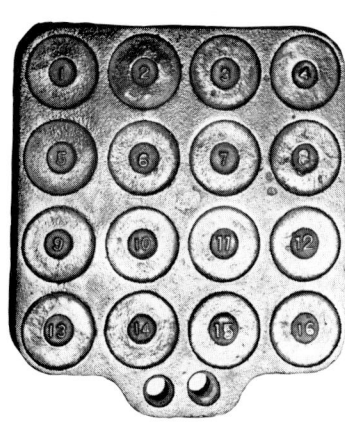

An assayer's cupel tray was used in one of the final processes in determining the mineral content of ore. The tray measures about eight by eight inches.

In the fall of 1848 Colonel Jonathan Stevenson mined at Mokelumne Hill with about a hundred of his men from the New York Volunteers who had been mustered out after the Mexican War. Colonel Stevenson wrote a code of mining laws for his men that is claimed to be the first such code to be drawn up in California.

It was here too that the first large nugget found in the Calaveras region was sighted. One of Stevenson's soldiers found it while he was getting a drink in the river, and news of this finding in 1848 stimulated the influx of gold seekers in 1849.

One man who mined this area was Captain Charles Weber, who organized the Stockton Mining Company as soon as he heard the news of Marshall's discovery of gold. Weber had already begun a settlement at Tuleburgh, or Stockton, at the head of navigation on the San Joaquin River. His party found gold in the sands of all the streams they explored up to Webber (*sic*) Creek near Placerville, where they mined successfully and gave Weber's name to the creek which, unfortunately, has always been misspelled on maps. Weber, however, did not mine for long. He was more concerned with the "golden" opportunity to develop his settlement as a supply base for miners.

Other early prospectors who left their names on the area were the Murphy brothers, John and Dan, who had been associated with Captain Weber in his explorations. In the fall of 1848 they led a party that mined on Coyote Creek, a tributary of the Stanislaus River, and found rich gravel at what became the first Murphys Camp. They discovered more rich gravel in the flat four miles to the north, and that site became Murphys New Diggings. The first Murphys Camp became Vallecito.

Henry P. Angel came at the same time and started mining at Angels Camp. James H. Carson went on south a few miles and found rich gravel in Carson Creek, at what became Carson Hill. These areas were rich in gold and attracted thousands of gold seekers in the great rush of 1849, when placer or free gold was so easy to obtain. Among the many sites along the Stanislaus River where gold was found was one place the early Mexican miners named Melones. The name was chosen because the thousands of flakes of gold discovered at this spot looked like melon seeds.

But the miners were searching for the Mother Lode, the source of the free gold in the streams, and it was finally located in the 1850s. The main lode of solid gold-bearing rock runs from Mariposa north along the foothills of the Sierra to Auburn in

Placer County. It runs across Calaveras County and Amador County in a northerly direction, and many famous mines tapped the ore in the years between the 1850s and the 1940s.

The main lode or veins crossed the Stanislaus River at Carson Hill and ran north through Angels Camp and past San Andreas into Amador County. There were also secondary veins or lodes called East Belts and West Belts to the Mother Lode.

In addition to these veins, some rich quartz or hard rock mines were developed. Carson Hill on the Stanislaus River was one of the richest concentrations of solid gold in hard rock in the Mother Lode. The first hard rock mine there was staked out in 1850 by John William Hance, who took in some partners to help him develop the mine. One of these partners was Colonel A. Morgan, and this mine became the famous Morgan Mine on Carson Hill.

Working their sluice box and gold rocker near Angel's Camp, Calaveras County, in the 1850s are Wade Johnston, standing, and Louis Weisbad and James Waters. Some of their descendants are still in the area today.

Mokelumne Hill, one of the oldest gold camps, traces its beginnings to the rich deposits found along the Mokelumne River. In the background is the famous "Jackson Buttes," in the center is the Leger Hotel, and to the left are the remains of the three-story Hemminghoffen and Suesdorff brewery. The walls of the brewery are no longer standing.

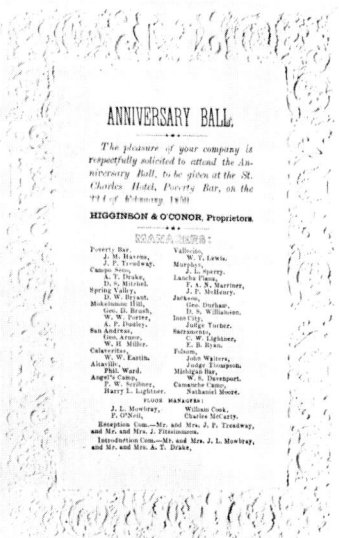 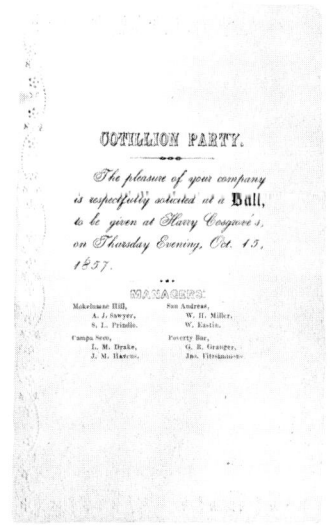 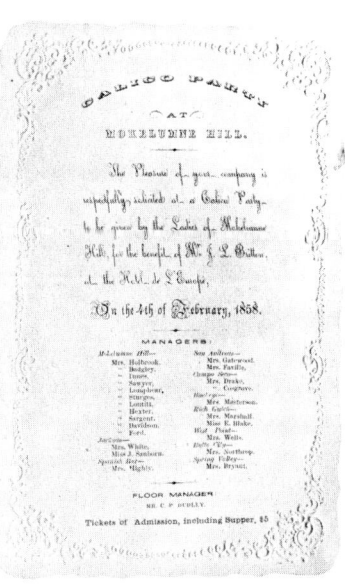

These are examples of social festivities in and around Mokelumne Hill in the 1850s.

Views of the Leger Hotel in Mokelumne Hill in the 1920s and the 1970s. A portion of the building was used as a courthouse from 1852 to 1866, when the county seat was moved to San Andreas. In 1874 the hotel was called the Grand. After a fire destroyed the ballroom in 1879, it was restored and its name changed to the Leger, after a Frenchman, George Leger.

In the late 1800s this building housed the telegraph office. Located south of the Hotel Leger, in later years it became Peter's Bakery, known as "Ye Old Moke-Hill Bakery." It is still in use today as Viola Perry's "What Not Shop."

A 1979 view of the Mokelumne Hill Community Church, which was started in 1856. It is claimed to be one of the oldest Protestant churches in continuous use in the state.

The Mokelumne Hill IOOF Hall with its iron shutters and doors was built in 1854, and the third story was added in 1861. It is supposed to have been the first three-story building in the Mother Lode. The first newspaper, the Calaveras Chronicle, *was published in this building; it also housed the Wells Fargo Express and Adams Express offices. Remarkably, the building still stands today.*

This hotel, now long gone, was at Fourth Crossing, between Angels Camp and San Andreas on Highway 49. The spot was the fourth crossing of the Calaveras River from Stockton to the mines, and it was a noted stopping place for the freighters hauling supplies to the gold fields.

The Guinn Mine, one of the larger producers, yielded over $7,000,000 in gold from 1883 to 1908 at a time when gold was valued at $20.67 an ounce.

Stockpiled lumber ready for shorings in the Guinn Mine, located five miles southwest of Mokelumne Hill in Rich Gulch.

Main Street in San Andreas in 1892, an election year.

The town of San Andreas as it was in about 1900. Established in 1849 by Mexican miners fourteen miles north of Angels Camp on Gulch Creek, the town was first located about one-fourth mile from its present location. It was named for Saint Andrew.

Gold scales on display in the old courthouse in San Andreas, now a county museum.

Except for the automobiles, Main Street in San Andreas still looks much the same as it did when this photograph was taken in 1925. This view is to the south.

The Stockton-Murphys stage stopped in front of the Metropolitan Hotel in San Andreas. It burned down in 1926 and was never rebuilt. Today there is a small park on the site, at Main and Court Streets.

The San Andreas Joss House, or Temple of Worship. Many of the mining camps had large settlements of Chinese miners. The Calaveras County High School is located on this site today. The little boy in the center of the picture is Emmett Joy, who was born in San Andreas in about 1900. He later became historian for the Grand Lodge of the Native Sons.

Saint Andrews Catholic Church was built in San Andreas in 1857. One hundred years later it was replaced by the present church.

The County Hospital in San Andreas was built in the early 1940s. It is still standing and in use.

Judge James Alexander Smith was Superior Court Judge of Calaveras County from 1917 until his retirement in 1957. He was an outstanding authority on Calaveras County history and shared his research in numerous articles published in local papers.

The San Andreas Government Center, located on the edge of town on the Calaveritas road, was completed in 1967 and houses all government offices.

Isaac Baker took this daguerreotype of the main street of Murphys looking west in the summer of 1853. This is the earliest known picture of Murphys; all of these buildings burned in the great fire of 1859.

These mixing tanks measure 150 feet across. The Calaveras Cement Company (now the Calaveras Cement Division of the Flintkote Company) also operates a forty-acre quarry near Vallecito. The company ships 6,000,000 tons of cement per year.

In 1922 William Mein and William MacNider became interested in the limestone deposits at the Kentucky House, a trading post two miles south of San Andreas, and in 1925 ground was broken for the Calaveras Cement Company. At its formal opening in 1926, the 15,000 guests present constituted the largest gathering ever in Calaveras County. That record has long been broken, with the Jumping Frog Jubilee, the Mt. Reba Ski Resort, flea markets, and rock concerts.

Cement from this plant has helped build the San Francisco-Oakland Bay Bridge, the facilities at Mare Island and Hunters Point naval yards, and the Delta-Mendota Canal.

Main Street in Murphys in about 1880. The two-story building on the left is the Sperry Hotel, which later became the Mitchler Hotel.

Taken in 1890, this photo shows stage time at the Mitchler Hotel in Murphys. Coaches stopped on the trip between Milton and Angels Camp and the Big Trees. In 1962 the hotel became known as the Murphys Hotel, after its purchase by the Ale and Quail Corporation, a group of Stockton investors.

Early view of the remains of once-rich Murphys Flats, then called "Murphys Diggings," behind the present Murphys Hotel. Gravel here was the richest of any deposits in Calaveras County, and claims were limited to eight square feet. Wells Fargo shipped over $16,000,000 in gold dust from their Murphys office between 1850 and 1860. It is said that John Murphy, the founder of the town, left with $1,500,000 in gold from the flats.

This is the North Ditch flume that carried domestic and general purpose water from Murphys to Angels Camp. The original flume was built by the Union Water Company in 1853 to serve placer mines. This picture was taken in 1978 along the Murphys Grade Road between Murphys and Altaville.

This panoramic view shows the Central Hill Hydraulic Mine, located one mile south of Murphys on an ancient riverbed that was later filled with a lava flow that extended to Vallecito. The water to operate the hydraulic monitors and sluice boxes here was brought by the North Ditch flume and was transported across Murphys Creek on a suspension flume 290 feet above the canyon floor. This spectacular engineering feat was completed on November 15, 1857.

Hydraulic mining in the Mother Lode.

An over-shot water wheel used for power to operate a gold mill near Murphys, circa 1860.

The two-cell Murphys jail was only a holding facility until lawbreakers could be transported to San Andreas. The concrete structure was built in 1913 by Frank Forrester, Price Williams and Pat Kaler, local carpenters. The first offender put into the jail was one of the builders, Kaler, who celebrated the completion of the job with a little too much to drink.

A ladies' club picnic at the Murphys South Ditch flume, which brought water from the Stanislaus River, fifteen miles away. The water was used for operating the Douglas Flat and Chio mines in Vallecito.

Harry Bray   Riley Senter   W. R. Senter   Bee Matteson

The P. L. Traver General Store in Murphys was built in 1856 and for years was known as the Riley Senter Store. It also housed the Wells Fargo office. In 1914 the building was converted into an auto repair garage which was in business there until the 1940s.

The old Senter building was purchased by Coke and Ethelyn Wood in 1949. They restored it and converted it into the "Old Timers Museum." Thousands of tourists have enjoyed the many old pictures and artifacts of the Mother Lode displayed here. The building is located in the center of Murphys.

Historic Murphys School, built in 1860, is now used as a Community Center. It has been replaced by the Dr. Albert Michelson School on Highway 4 nearby. Dr. Michelson, Murphys School's most famous graduate, was a Nobel Prize winner.

The First Congregational Church in Murphys was built in 1895 and still serves more than one hundred members.

Douglas Flat School is believed to be the oldest one-room schoolhouse in Calaveras County and possibly in California. It is located on Highway 4 between Vallecito and Murphys. The gravel in the playground was gold-bearing, and one of the inducements used to persuade a teacher to take this school in the early years was that he or she could wash the gravel during recess and lunch periods. Douglas Flat residents claim that the school dates to 1854.

A typical sixteen-horse freight wagon at the chlorination works of the Utica mine, about 1890, prepared to make its run between Angels Camp and Murphys.

Freight wagons like these were a daily sight on the roads from Stockton to the Southern Mines. They had a load capacity of 2,000 to 5,000 pounds. Note that there is no seat for the driver; he rode one of the "wheel horses" or walked.

The Utica mine, first located in the early 1850s, was purchased in 1865 by Judge Delos Lake and named after the Judge's home town in New York. During the 1890s the mine was extremely prosperous; in one month in 1895 it yielded $400,000 in gold bars. At the height of its success its owners reported a net income of $3,000 a day. Total production of the Utica mine is estimated at $17,000,000.

Angels Camp, the only incorporated town in Calaveras County, is shown in this 1979 view, looking south on Main Street. The town was started in 1848 when Henry Angel discovered red gold in the gravels of what is now Angels Creek. He set up a trading post, but the gravels were not rich, and it was not until the main vein of the Mother Lode was discovered in the 1850s that Angels Camp became an important mining center.

These cars were used to transport gold-bearing ore out of mine tunnels. They were either pulled by mules or pushed by hand.

This view shows the Angels mine which produced over $3,500,000.

The small Angels Mill mine located on the north side of Angels Camp produced $2,000,000 in gold in combination with the Angels quartz mining company. The mine closed in 1918.

In 1900 Edward True and Joe Nava posed for this picture with the oxen of their logging wagon. Twelve to fourteen horses were used in front of the oxen in hauling these large logs to the Manuel Lumber Company near Murphys.

The logging team of Gus Emerald, driver, and Bill Stockton, on their way from Murphys to Angels Camp with mine timber.

The steam traction engine, as it was called, pulled five wagons loaded with 40,000 feet of lumber. It never became popular for haulage on public roads because of its damaging effects on road beds and bridges, but it was a forerunner of the track-laying tractor. The lumber is from the Manuel Lumber Company near Murphys.

The Best steam traction engine was built in San Leandro, California about 1900 and was used to haul lumber for the Manuel Lumber Company of Murphys and Raggio Brothers of Angels Camp. Dubbed "Ol' Beth," the engine is now part of the exhibits at the Calaveras County Museum at the edge of Angels Camp.

This is a 1977 view of Vallecito, which means little valley. Its name was changed from Murphys Old Diggings by Mexican miners in 1854.

The bell atop the quartz rock monument in front of the Vallecito Community Church originally came from New York in 1853. It was purchased with funds contributed by local residents and was placed in a large oak tree. Until a severe wind blew the old tree down in 1939, the bell was used to summon people for many reasons. This stone monument was erected later in 1939 by the Grand Parlor of the Native Sons of the Golden West.

The Dinkelspiel Store and Wells Fargo Express building, built in 1854, is the most significant historical structure surviving in Vallecito.

The ruins of the residence of James Romaggi, built in the 1850s, are a short distance from Carson Hill on Highway 49, at what was known as Albany Flat. Romaggi came to Calaveras County from Genoa in 1850; he preferred farming to gold mining.

Carson Hill, located south of Angels Camp on Highway 49. In the background is the famous "Glory Hole" which in 1854 produced the largest mass of gold ever found in the United States. When it was placed upon Toll's Express scales in Stockton, it weighed 195 pounds and was worth $43,534. The Carson Hill mine was the largest producer of gold in the area between 1850 and 1942, yielding an estimated $26,000,000.

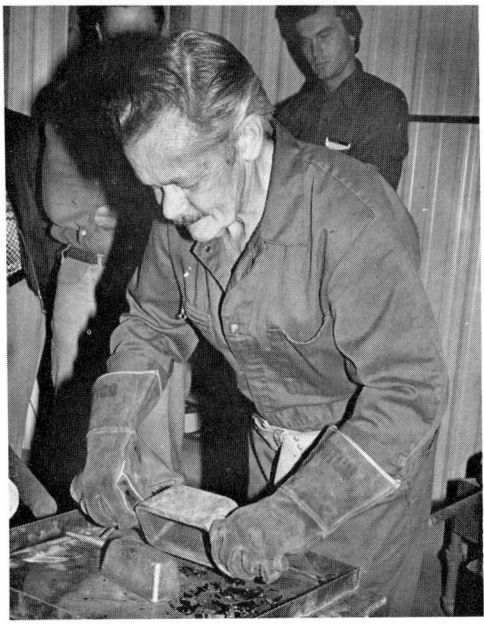

Troy Gold Industries Ltd. of Calgary, Alberta, Canada, has announced the pouring of the first gold produced at its Blazing Star Mine.

The mine, first worked in the 1890s and dormant throughout most of this century, was reactivated by Troy Gold last year after extensive testing. It is now on a twenty-four-hour, seven-day production schedule, turning out approximately one hundred fifty tons of ore per day.

While test drillings have revealed ore values as high as nineteen ounces of gold per ton, the overall average is expected to be in excess of one ounce, still considerably higher than that of the principal Canadian and South African mines.

Ore from the Blazing Star has also been found to contain recoverable amounts of silver and tungsten, which by themselves should pay the cost of mining and milling, say company officials.

The mine, located fifty-five miles southeast of Sacramento in the heart of California's Gold Rush country, is a highly efficient ball mill operation. After the gold, silver, and tungsten are extracted through flotation and filtration, the remaining sludge has been found to contain virtually no trace of valuable metal.

Troy Gold is not using the old original mine shaft except for ventilation and emergency exit purposes. Instead, it has sunk a spiral ramp that descends at a gradual angle, crisscrossing the various veins. The shaft is now at the 420-foot level.

California's Mother Lode back in production. Al Farrell, president of Troy Gold Industries Ltd., displays the first solid gold ingot produced at the company's Blazing Star Mine in West Point, Calaveras County. The bar weighs in at 200 ounces, making it worth more than $62,000 in today's market, where gold has soared above $310 per ounce. Troy Gold began reactivating the old Mother Lode mine in 1977 and is now using modern technology to extract substantial amounts of precious metal left untouched during the great California Gold Rush.

Old Tecumseh, a colorful Miwok Indian at Sheep Ranch, was known as the jolliest man in the Mother Lode.

Several Miwok Indian roundhouses stand on the hill north of Murphys, on the Sheep Ranch road. The last rancheria in this area burned in a brush fire in 1927, and the Indians re-established at Sheep Ranch.

Sheep Ranch women and children on an outing posed on and around the mill wheel of the Chavanne gold quartz mine, circa *1900.*

The main street of Mountain Ranch, about 1920. The smallest post office building in the United States, six by eight feet, was on this street at one time. It is now on display at the Pollardville attraction in Stockton. Mountain Ranch was once a mining camp and is located about ten miles east of San Andreas.

The upper story of the Anderson Hotel at Sheep Ranch was constructed near Mountain Ranch as a one-story building and then moved to Sheep Ranch where it served as a general store and living quarters. In 1904 it was raised and the present ground floor was built underneath it. There were eighteen bedrooms upstairs, and downstairs were some bedrooms, a dining room, kitchen, and bar. In the background is the Sheep Ranch mine.

A group of women enjoy a game of croquet in Copperopolis outside the Meador home, which has long since disappeared.

Memorial services for Abraham Lincoln were held in the Congregational Church in Copperopolis in 1865, a year after it was completed. Since this picture was taken, the building has been restored and is now used by the Community Club.

Armory Hall in Copperopolis was built in 1862 for the Union Guards, who protected the important copper mines from possible saboteurs during the Civil War. Copper was discovered near here in 1861, just in time to provide vast quantities of the strategic metal for the Union forces. It was used in making bronze for cannons, fastenings for harnesses, and munitions, among other things.

The third Catholic Church built in Copperpolis was erected in 1916 and torn down in 1970.

The main street in Copperpolis, about 1890. The Union Hotel is on the left. The buggy on the right is in front of the building now known as The Corner. At one time Copperpolis had a population of more than ten thousand.

Ore from the Napoleon copper mine, southeast of Milton, being unloaded on September 10, 1907. The ore wagons were pulled to the loading dock by the Best steam traction engine.

Milton, shown here about 1890, developed because it was the terminus of the Stockton and Copperopolis Railroad. The line was to continue to Copperopolis, but financial difficulties prevented completion. A huge celebration was held when the first passenger train pulled into Milton on July 4, 1871. In 1888 the railroad was taken over by the Southern Pacific Railroad Company, which operated it until it was discontinued in 1940.

In a rich placer area, Jenny Lind was a busy supply center in 1849. The following year it had a population of nearly five hundred, mostly of Mexican and Chinese descent. Although the town has the same name as the famous "Swedish nightingale," historians say that there is no connection. Founded by Dr. John Lind, Jenny Lind is located about six miles north of Milton, just off Highway 26. The large hole in the foreground was caused by hydraulic mining.

The gold dredge "Calaveras" operated at Jenny Lind. It was the most successful of four dredges which operated in the area until about 1930. While gravel piles from dredging are much in evidence today, there is little left of the old town.

Southern Pacific locomotive #1025 with passenger coach and car of the old San Joaquin and Sierra Nevada narrow-gauge railroad, at Valley Springs in 1888. The same year Southern Pacific purchased the twenty-two-mile line, which had been completed in 1885 to serve copper mines in the Valley Springs district.

Carriages await their passengers at the end of the railroad line at Valley Springs. In 1928 the line was extended to the cement plant near San Andreas.

This 1979 photograph shows the Alexander Wheat house in Double Springs which was built in the 1860s of sandstone taken from a quarry on his ranch. Wheat was an important Calaveras County rancher who also served in the state legislature. The home is now occupied by his granddaughter, Mrs. Sadie Hunt.

This ramshackle building at Double Springs was the first Calaveras County Courthouse. It was prefabricated in China in 1849 and imported to California. Note the door-like panels for its sides. A protective building was built over the old landmark in 1950 by the Calaveras Grange and efforts made to preserve it.

The Immaculate Conception Catholic Church and schoolyard at Camanche in the 1950s, now deep below the waters of Camanche Reservoir. There were several brick, stone, and wooden buildings in the town dating back to the days when it was a mining center known as Clay's Bar; all were completely torn down before water covered the site. This photograph was found among the rubble when the townsite was being cleared. A replica of the church has been constructed at Lake Camanche Village. The dam was completed in December 1963.

The Snider Lumber Company, located at Wallace on Highway 12, between Clements and Valley Springs. The company produces unfinished lumber cut into two- by twelve-inch planks which are shipped to the finishing mills. It is the most completely automated mill in the state. Wallace is also noted for fine olive groves. The olives are shipped to the "olive capital of the world," Lindsay, California, for processing.

Pardee Dam forms a reservoir with a water surface of 2,257 acres. An 18,750 KVA power plant at the dam generates power which is sold to the Pacific Gas and Electric Company. Pardee Dam and Reservoir was opened to the public in 1958. Facilities include a launching ramp, overnight camping, and trailer parking.

Camanche Dam, completed in 1964, is 2,640 feet long and 171 feet high. It is located about ten miles downstream from Pardee Dam and Reservoir. The lake created by Camanche Dam has a surface area of 7,622 acres and a shoreline of sixty-three miles. It is open all year and has extensive facilities for day and overnight use.

This beautiful Victorian house was started in 1857 as the residence of Ben Thorne. It has been preserved and is a showplace in San Andreas.

Ben Thorne, sheriff of Calaveras County for almost forty years, helped in the capture and conviction of Black Bart.

Don Cuneo and associates established a motel and restaurant in San Andreas in the 1960s and named it for the fabled bandit, Black Bart. It is located across the street from the courthouse where Bart was sentenced.

"CLAMPER WALL"
In Murphys

The *E. Clampus Vitus* "Wall of Comparative Ovations," consisting of sixty-five beautifully sculptured relief plaques of famous Clampers, historical characters, and events in the West, is a delightful and interesting display.

The wall is located in Murphys, an engaging little mining town on the East Belt of the Mother Lode, on an old stone wall of the Old Timers Museum. The sculptor is William Gordon Huff, one of California's best, and also an enthusiastic Clamper. An interesting half hour can be spent reading the captions on such plaques as those to the Dying Buffalo, End of the Trail, Chief Truckee, Saber-toothed Tiger (fossil animal of California), Pony Express Rider, Sir Francis Drake, John Murphy, a Central Pacific locomotive, great patrons of *E. Clampus Vitus* and famous Clampers.

The ancient and honorable order of *E. Clampus Vitus*, the story goes, was organized in Mokelumne Hill by Joe Zumwalt in 1851 and spread to all the gold camps of the Mother Lode. It was a fun and welfare order with a "secret" ritual that was a parody on all fraternal orders. Emphasis was placed on the initiation of candidates for membership, who were charge a big fee for an invitation to buy the liquor and food. The Clampers' prime concern was the welfare of widows and orphans.

As the mining camps, with their rough turbulent life, emptied, E.C.V. lost its vigor and practically died out. In 1931 Carl Wheat, Ezra Dane, and Leon Whitsell, three author-historians on a visit to Columbia and the Mother Lode, agreed to try to revive the order. Their efforts were met with hearty enthusiasm, and chapters were gradually organized in San Francisco, Sacramento, Stockton, Virginia City, Nevada, and throughout the gold country.

In the 1950s a Grand Council of all thirty chapters was organized, and since then delegates have met to celebrate Memorial Day weekend at Murphys, wearing their colorful red shirts and black hats. Emphasis in the revived order is on preserving history as well as having fun, and over two hundred historical plaques have been dedicated by different chapters in California and Nevada.

The E.C.V. "Wall of Comparative Ovations" in Murphys is a "holy" place for all Clampers, and fascinating for anyone.

The noted historian Coke Wood ("Mr. California") and Hal Goodyear, ex-Sublime Noble Grand Humbug of the Grand Council of E Clampus Vitus, at the dedication of the plaque to Archie Stevenot by E.C.V., on May 29, 1977.

The Wall of Comparative Ovations of E Clampus Vitus on the side of the Old Timers Museum in Murphys, Calaveras County.

V

Tourist Attractions Of Calaveras

The yellow metal from the mines is not the only kind of gold produced in Calaveras over the years.

The tourist gold from the visitors who come to see the "wonders of Calaveras" has been very important to the county's economy from the beginning. The Big Trees, limestone caves, beautiful mountain scenery, hunting, fishing and skiing have each brought many visitors to the area.

The Calaveras Grove of Sierra Redwoods (*Sequoia Gigantea*) were the first of the giant trees to be discovered and publicized. It was in 1852 that A. T. Dowd, a hunter for the Union Water Company, was following a wounded bear and first saw these enormous trees. He was there investigating the area as a potential site for the aqueduct from the Stanislaus River to Murphys Diggings that the Union Water Company planned to build.

In 1853 Captain James Hanford and a crew of five men took twenty-two days to cut down one of the trees in order to determine its age. No saw was adequate, so augers were used to sever the trunk, but the tree did not topple until a wind came up. By counting the rings they determined that the tree was over 1,500 years old. A building was later erected on the stump of this tree. The building was used for various purposes for many years and quite possibly was the site of the first bowling alley in California.

In 1854 George Gale and a crew of men stripped the bark from the "Mother of the Forest" up to 116 feet in order to reconstruct it and prove its size to the skeptical Easterners in New York and Boston. Subsequently, in 1857, the bark of the tree was shipped to the Crystal Palace in London, where it was admired until 1866, when a fire destroyed it. The old dead tree is still standing, badly burned but still called "Mother of the Forest."

In 1854 W. W. Lapham and A. S. Haynes built a hotel at the Calaveras Grove to accommodate scientists and other visitors who came to see the magnificent trees. James L. Sperry and John Perry purchased the grove in 1858, and after the first hotel burned in 1864, Sperry built the larger Big Trees Hotel among the trees. John Perry, who had acquired a hotel in Murphys, had died by this time and so was never involved in the Big Trees Hotel. Most guests there as well as at Perry's Murphys Hotel, fifteen miles down the Ebbetts Pass Road, have their names preserved in the old registers.

The North Grove of the Calaveras Big Trees became part of the California State Parks System in 1931, thanks to the efforts of such men as Desire Fricot of Fricot City. The South Grove, about two hundred large trees south of the Stanislaus River, also became part of the State Parks System, on September 9, 1961. This grove is only accessible by trail from the North Grove Road. Many individuals and organizations are responsible for bringing these groves into the State Parks System, but much of the credit goes to Stuart Gibbons and Mrs. Owen Bradley of the Calaveras Grove Association, who worked for years to raise funds and rally support.

These trees are known to be between two thousand and four thousand years old. They attract thousands of visitors to Calaveras County each year, making them an economic boon as well as a scenic wonder.

The Limestone Caves have long been an important part of Calaveras history. The first one to be famous was at Cave City, located near Mountain Ranch. As its fame spread during the 1850s and 1860s, many curious people came to visit the cave, and Cave City became a thriving little village, supporting a hotel and several stores. Over the years the cave was neglected and damaged by floods, and it lost its appeal as a tourist attraction. The Mercer's Caverns at Murphys and the Moaning Cave at Vallecito, however, continue to attract many visitors each year.

Mercer's Caverns is named for Walter J. Mercer, a miner in Murphys Camp, who discovered it in September 1885 on his way home. One mile north of Murphys, on San Domingo Road, Mercer noticed cold air coming from a small opening in the rocks. He borrowed tools and rope from the nearby Oro y Plata Mine and found a large cavern about thirty feet down. Now called the Gothic Chamber, this cavern has a ceiling covered with large masses of stalactites. With additional help Mercer explored the cavern as far as ninety feet below the opening and discovered beautiful formations such as the Organ Loft, the Angels Wings, and many others.

In June 1887 the caverns were opened for public tours. Candles, torches and kerosene lamps were used for lighting. Electricity was installed in 1901. With its great variety of stalactite and stalagmite formations that took millions of years to form now beautifully lit with colored lights, Mercer's Caverns is a popular stop for visitors to the Murphys area.

Another important limestone cave, located near Vallecito, is Moaning Cave, first explored by gold miners in 1851. Scientists have researched and excavated the cave and found human remains which they estimate to be at least twelve thousand years old. The bones have been preserved by the deposits of mineral-bearing waters. The visitor descends a steel spiral staircase 100 feet to the floor of the main chamber, where the major formations are located. Moaning Cave was so named because an early visitor thought the air currents sounded like a moan. It is located on the Parrots Ferry Road between Vallecito and Columbia State Park.

Very near this cave, on Coyote Creek Canyon, is the Natural Bridges, one of the natural wonders of this area. With the help of an old miner and recluse named "Batch" Barnes, who lived underneath the limestone bridges, the area became quite a popular attraction. Old Barnes eked out a living by mining the gravel of Coyote Creek and would entertain visitors with stories about the bridges, which had been formed over the centuries by the creek's overflow. Unfortunately, the land where these bridges are located is now privately owned, so they are not open to the public except by special permission.

Probably the greatest tourist attraction in the Calaveras area is the Jumping Frog Jubilee, held at Angels Camp each year since 1928. In that year the Boosters' Club of Angels Camp wanted to celebrate the paving of the town's main street and planned a western-type celebration. Someone who had read Mark Twain's "Celebrated Jumping Frog of Calaveras County" proposed that they sponsor a frog jump and call it a "Jumping Frog Jubilee." The top prize would be awarded for the frog that jumped the farthest.

The Boosters' Club and Angels Camp itself were overwhelmed by the large crowd that came to see their curious exhibition. The merchants, of course, were delighted, and plans were made for an annual event. After three years the celebration was moved two miles south to "Frogtown," and in 1935 it was combined with the

County Fair, held annually during the third weekend of May.

Tremendous crowds come from near and far to see the frogs jump. The four-day Jumping Frog Jubilee is a prime source of tourist gold—and a great deal of fun!

A more recent but also significant tourist attraction is the Bear Valley and Mt. Reba skiing and resort complex. Although it is located in Alpine County, just over the Calaveras County line, access during the skiing season is from Highway 4 through Calaveras County.

C. Bruce Orvis is the person most responsible for the Mt. Reba Ski Resort and the Bear Valley commercial and residential development. The area was part of the Stanislaus National Forest until Orvis was able to acquire it in a land trade with the United States Government. He succeeded in interesting a group of businessmen in his idea of a ski and recreation center and formed Orvis and Associates. In May 1966 they obtained a special use permit, and by July 1967 construction was under way.

The opening of the Mt. Reba ski area at Bear Valley on December 15, 1967 was attended by a large and enthusiastic crowd. Three double chair lifts were available for skiers in 1967. The resort has become so popular that eight double chair lifts were in operation by the 1977-78 season. Appropriate slopes have been prepared for all levels of ability, involving about 800 acres on the slopes of Mt. Reba. Two separate firms are involved in this development at present: Orvis' Bear Valley Company and the Mt. Reba Corporation, led by Dennis Rasmussen.

This was one of the first automobiles to cross the Big Trees toll road at Dorrington, which extended across Ebbetts Pass to Silver Mountain. Circa 1911.

One of the largest trees in Calaveras Big Trees State Park is the "Mother of the Forest," shown in the center of this picture.

According to the nameplate in the upper lefthand corner, these visitors are encircling the Washington, one of the Sequoia Gigantea *redwoods at Calaveras Big Trees, in this 1880 picture.*

Mercer's Caverns, extending ninety feet below the surface, was opened to visitors in 1887. It is one of fifteen caves in North America to have deposits of flosterns. This aragonite formation resembles coral and is also known as the iron flower. The example in this picture is known as the "Diamond Cascade."

The stump house in the north grove of the Calaveras Big Trees. The tree was cut in 1853 and the stump was used for many things, including a dance floor, restaurant, newspaper office, and church services. It was crushed by snow in 1939, and the State of California never replaced it. The tree trunk was twenty-five feet in diameter and weighed 240 tons; it was estimated to be 3,000 years old. The stump remains and is the leading tourist attraction in the Park.

"Batch" Barnes poses with a group of visitors underneath his home, the natural bridges in Coyote Creek Canyon near Vallecito.

"Batch" Barnes in the outfit he wore for a parade in San Francisco in 1890.

The Calaveras Big Trees Hotel, seen through the "Sentinels" from the carriage road that leads to the Grove of Big Trees.

James L. Sperry, who built the Big Trees Hotel in 1864, is shown with his wife, Mehitable.

This close-up shows the Big Trees Hotel in the 1890s. It burned in 1943. During its existence the hotel accommodated famous visitors from all over the world, including Thomas A. Lipton, the tea king, the Rothschilds from Europe, Mark Twain, Ulysses S. Grant, and many others.

Mark Twain received the inspiration for "The Celebrated Jumping Frog of Calaveras County" soon after his arrival in the Mother Lode in December 1864. The story was warmly accepted by a nation recovering from war, and it helped make him famous.

Main Street in Angels Camp in the 1920s, before the street had been paved and before the Jumping Frog Jubilee had become an annual event.

The Angels Hotel circa 1890. This is where Mark Twain heard the story that gave him the idea for "The Celebrated Jumping Frog of Calaveras County." Today the old building is used as an auto supply company and a general store.

Angels Hotel, 1979.

117

One of several events at the annual Jumping Frog Jubilee is the Children's Parade, shown here in this 1940s photo (upper left). Notice the wash hung out to dry overhead.

A champion jumper with his prize money, upper right.

Lower left, Roy Weimer shows off "Lucky," the 1954 World Champion with a jump of sixteen feet ten inches.

An unidentified youngster (bottom right) appears to be showing his frog exactly how it should be done.

More than ten thousand people visit the Mt. Reba ski area over a holiday weekend. It is estimated that the population of Calaveras County doubles on these occasions.

VI

Tuolumne County

Tuolumne County lies roughly between the Tuolumne and Stanislaus Rivers and between two other Mother Lode counties, Mariposa and Calaveras. It may be called the heart of the Southern Mines, a term first used to describe the rich placer or "free-gold" mines in the area later to become Amador, Calaveras, Tuolumne and Mariposa Counties.

The first gold in Tuolumne County was discovered on Woods' Creek near Jamestown in the summer of 1848 by a party led by Reverend James Wood. They found the gold so plentiful that they could make $200 to $300 a day with only a pick and a knife. Soon other miners were crowding into the diggings of this rich little creek, and Colonel George James founded the settlement of Jamestown, popularly called "Jimtown."

As the word of rich gravel in the streams of this area spread, many Californians and Mexicans came into the area. By the fall of 1848 news of this rich gold mining had reached Sonora, Mexico, and large numbers of Mexican miners joined the trek to the gold country. Some of these gold seekers settled near Jamestown, and this settlement was called the Sonorian Camp. The Rush of 1849 brought many more miners to wash the gravel of the flats and bars along the streams, especially the Tuolumne and Stanislaus Rivers.

The Tuolumne area became a part of the San Joaquin District when Governor Bennett Riley called for an election in ten districts in California to choose delegates to meet in Monterey in August 1849 to draw up a state constitution. On November 13, 1849, the voters overwhelmingly approved the constitution, and state government began in California in December 1849, before official ratification from Washington.

In February 1850 Tuolumne became one of the twenty-seven original counties created by the Legislature. General Mariano Vallejo, a member of the Senate from Sonoma, was the authority on the names of several counties. He explained that *Tuolumne* was the corruption of a Miwok Indian word *Talmalamne*, which means "stone wigwams." Other authorities say the word has two parts, *umne*, meaning "people of," and *tual*, meaning "stone houses." The name Tuolumne was given to the river, the county, and the settlement.

By 1850 the largest and most important settlement in Tuolumne County was Sonora. It became the county seat, and county officials were elected in the spring of 1850. The first courthouse was built in 1853.

The area was originally settled because of the placer gold in the flats, creeks and rivers, and Tuolumne County became the center of the Southern Mines. However, it was soon discovered that the quartz veins of the Mother Lode, the source of the placer gold, passed through the center of the county, with its East Belts and West Belts, and many quartz or hard rock mines were developed. Among them were Eagle-Shawmut, App, Dutch, Sweeney, Juniper, Rawhide, Harvard, and Soulsby.

The miners also soon realized that ancient lava flows followed river beds and rich placer gold was beneath them. To the west of Jamestown is Table Mountain, an ancient lava mass some forty miles long. The mines were tunneled in honeycomb fashion underneath most of the mountain. These mines have all been closed now, but one of them, the Humbug, produced about $4,000,000 in gold during its lifetime.

The Big Bonanza Mine in Sonora produced a great deal of gold and is said to be the biggest pocket mine ever found in the Mother Lode. It was located on Piety Hill in the center of present-day Sonora. Because of its central location among such mining areas as Columbia, Squabbletown, Sawmill Flat and Yankee Hill, Sonora became a commercial center as well as a mining town. This gave it a permanence that other diggings lacked. Sonora became the "Queen of the Southern Mines," one of the wildest towns in the gold country. Its main thoroughfare, Washington Street, was lined with all kinds of buildings, but most of them were destroyed in a series of fires.

Sonora, or the Sonorian Camp, was the first large mining settlement created by the gold rush, but it soon had competition from booming Columbia, founded in 1850 by two brothers, Thaddeus and George Hildreth. In March 1850 John Walker, a member of the Hildreth prospecting party, discovered gold here. The camp was first called Hildreth's Diggings and then American Camp, but it was named Columbia by the time of its incorporation in 1854. Columbia developed rapidly and was known as the "Gem of the Southern Mines." By about 1880 rich, easy mining nearby had produced about $87,000,000 in gold. Because of the surface mining, large numbers of gold seekers rushed into the mines, but they needed water to wash the gravel. The Tuolumne County Water Company was formed in June 1851 to bring water into the area through a network of reservoirs, flumes and ditches. Miners formed the Columbia and Stanislaus River Water Company in 1854 and by 1858 completed a sixty-mile aqueduct to bring water in from the Stanislaus River.

The first tents and shanties were gradually replaced by more permanent wooden buildings. By 1852 streets were being laid out, and Columbia had more than one hundred and fifty stores, shops and saloons. The first big fire, in 1854, burned the entire center of town except for one brick building. After that, many brick structures were built, with iron door and window shutters for fire protection, and these were all that survived when another fire in 1857 destroyed all structures made of other materials. Rebuilding began immediately, with fireproof materials specified. At this time a volunteer fire department was organized, and in 1859 the town acquired the famous little handpumper, affectionately called Papeete.

After 1860 most of the easy surface mining was exhausted, and the population began to decline from around ten thousand in the 1860s to about five hundred in the 1870s and 1880s. However, because Columbia always maintained some commercial activity and remained one of the best preserved gold towns of the Mother Lode, a movement was started to preserve it as a state park. A bill creating the Columbia Historic State Park was passed by the California Legislature and signed by Governor Earl Warren on July 15, 1945. The State Park Service set about acquiring, preserving and restoring buildings as fast as funds were available. Today the old mining community is one of the finest interpretive state parks in the state, portraying life in a gold mining town of the 1860s and 1870s by restoration and exhibits.

The "gold" that makes Columbia, Sonora, and all of Tuolumne County prosperous today is from the tourists who visit the area in large numbers. More people walk the streets of Columbia during special events than ever lived there during the mining period. There are an estimated five hundred thousand visitors each year.

One of Columbia's main attractions is the summer theater at the old Fallon House, where the University of the Pacific trains drama students in all phases of dramatic production. Owen Fallon built the hotel in 1857 and later added a dance hall that became very popular in the community. His son James added a stage where plays were presented. For many years, while Columbia was the "Gem of the Southern Mines," the Fallon House was the center of social activity there, but when business in the town declined the building was neglected and it deteriorated.

Robert Burns of the College of the Pacific (later its great president) learned the Fallon House was in financial jeopardy and was for sale, and he raised funds to purchase it for Pacific in 1945. He saw its possibilities as a summer repertory theater for training drama students. With the cooperation of Professor DeMarcus Brown, head of the Drama Department, this dream became a reality and in 1949 the college performed its first summer season of plays to help celebrate California's Gold Rush Centennial.

The theater was deeded to the state in 1947 with the agreement that Pacific would be permitted to train its drama students there by presenting a summer theater program. The twenty-ninth successful season, with the largest attendance on record (over twelve thousand people), was held in 1978, under the direction of Dr. Sy Kahn, who took over from Professor Brown in 1969. The company of thirty-six people, including twenty student actors, presented forty-three performances of five plays in repertory. This company was responsible for the total operation of the theater.

In 1961 Eagle Cottage was built nearby as a dormitory and dining room for twenty students and six staff members. A joint effort of the University of the Pacific and the State of California, the cottage was constructed from a drawing of an old structure which had long since disappeared.

The old mining town that began a new career as a state park in 1945 also became the site of one of the most beautiful community college campuses. In 1966 the voters of the Modesto Junior College District voted overwhelmingly to enlarge their district to include all of Stanislaus and Tuolumne Counties as well as parts of several other counties. The enlarged area was renamed the Yosemite Junior College District. To provide higher education to the remote areas in the mountains, the trustees authorized the formation of a second campus in Tuolumne County, Columbia Junior College, located near Columbia State Park. This uniquely beautiful college is located among 160 acres of pine forest surrounding San Diego Reservoir, with the lake as the center of the campus.

Classes at Columbia Junior College began in the fall of 1968, before the campus was built. The college opened in rented facilities with 300 full-time students, a total of 1,140 part-time students, and a faculty of 15 instructors under the leadership of President Harvey Rhodes. The official dedication of the campus was held on May 17, 1970.

Since that time the campus has been ninety percent completed, and the expanded faculty offers a great variety of courses preparing students both for advanced college training and for careers in the local community.

The Westside and Cherry Valley Railway Theme Park is a new tourist attraction near Tuolumne.

In 1976 Glenn W. Bell, Jr. bought out the Westside Lumber Company mill property, for the purpose of establishing an amusement park similar to Knotts Berry Farm or Disneyland, but with a character of its own. Bell is famous for developing the 900 Taco Bell restaurants throughout the nation.

The 340-acre Westside and Cherry Valley Railway Amusement Park is called a living museum because it offers rides on Westside's restored 4½-mile narrow-guage steam railroad.

Tuolumne is a small lumbering town of some three thousand, located eight miles east of Sonora, just off Highway 108, the Sonora Pass Highway. For at least twenty years it has been a depressed community, hard hit by a strike and fire that destroyed the Westside Lumber Company mill. Bell saw the mill advertised for sale in the *Wall Street Journal* and purchased the entire property, including the narrow-gauge railroad and the mill pond, for $300,000. The pond became a twenty-five-acre lake.

The park opened in September 1978 after two years' work, but it will not be completed until 1990. At that time, Bell has told the Board of Supervisors, it should have an annual payroll of $5,000,000 and attract a million visitors annually, who will generate an expected $100,000 per year in sales tax.

Since it opened, Westside Park has averaged five hundred visitors each weekend and is thus an important part of Tuolumne County's economy.

Looking east along State Highway 120, about one mile west of Jamestown, in 1979. In the highway cut can be seen a large quartz vein, the formation in which hard rock miners often find gold. At the bottom of the grade is Woods' Creek, where gold was first discovered in Tuolumne County. In 1848 a seventy-five-pound nugget was also found in the same area.

One of the loveliest available views of Jamestown, which was founded in 1848 by Colonel George James, a San Francisco lawyer. There were a hundred flags flying from restaurants, taverns, stores and gambling houses following the discovery of gold along Woods' Creek.

Jamestown in 1979. The town was almost wiped out by a 1966 fire that destroyed most of the original buildings. The Willow Hotel, seen on the left, was restored and returned to a nineteenth century look in 1975. Three years later it burned again. Another rebuilding is planned, with as much as possible of the surviving structure to be used. President William McKinley is said to have been a guest of the hotel in 1901.

The Soulsby Mine became the first major hard rock gold mine in the county in 1855. In 1858, 499 miners were brought from Cornwall, England; they were the first English hard rock miners. By 1900 the Soulsby Mine had produced more than $7,000,000 in gold.

Jamestown Depot, called Railtown, is just off Highway 108 and features novelty shops, a restaurant, and antique shops. Three steam trains carry passengers on weekends and provide meals and entertainment. This portion of the Sierra Railway Depot, dating from 1897, was remodeled after it was completely burned. There was another tragic fire in November 1978, but there are no plans to rebuild the depot. The historic railroad shops nearby were undamaged.

These movie stills and memorabilia were among the items destroyed in November 1978 when the Jamestown Depot Museum burned. The museum was located on the second floor of the old Sierra Railroad terminal. More than one hundred and fifty movies and television series were filmed in the Jamestown area, including High Noon, Duel in the Sun, Union Pacific, *"Petticoat Junction" and "The Wild, Wild West."*

Washington Street in Sonora, 1890.

Early day view of Washington Street in Sonora.

A present day view of Sonora. The town's rapid growth has made it difficult to keep abreast of the needs for housing, natural gas, water, and parking spaces, but the inconveniences caused by these shortages is only temporary.

The main street of Sonora during the 1920s, looking south from the center of town toward the gateway to the Sonora Pass resorts.

Tuolumne's County's first courthouse was built in 1853.

The Masonic Lodge laid the cornerstone for the second courthouse on September 26, 1898 in Sonora.

The courthouse built in 1898 is still in use today.

M. J. Curtin ("Buchanan Mike") stands with his team in front of Pat Burk's blacksmith shop on Washington Street, circa 1890. The Sonora Theater was later on this site. Curtin was bound for the Buchanan Mine, twenty-eight miles southeast of Sonora, which was to yield more than $2,000,000 in gold.

Sierra Railroad Station at Sonora, right, and County Hospital, left, located at the south end of Washington Street.

The City Hotel was built in 1852 and still stands on Washington Street. The building now serves as the Sonora City Hall. The City Hotel and the Victoria Hotel were Sonora's leading hostelries.

A turn of the century view of Sonora, which got its name from the state of Sonora in Mexico. The town government was formed in 1849, and by autumn of the first year the population neared five thousand. The famous Holden Chispa Nugget, weighing over twenty-eight pounds, was found within the city limits, in Holden's Gardens.

The Sonora Herald *was published here from July 1850 to June 1859. The first issue was actually printed in Stockton, and the type was that of the* Stockton Times *and* Tuolumne City Intelligencer. *The press used was the old Ramage press brought to California in 1836 by Augustin Zamorano, which produced the first printing of any kind in California and, in 1846, the state's first newspaper, the Monterey* Californian.

The Sonora Herald *later sold the press to the* Columbia Star, *the first newspaper there. The story of this pioneer press ends here, where vandals burned it in a dispute.*

This wood frame and adobe building has served many owners and many purposes since it was constructed. Its major uses were as the Gunn House Hotel from 1851 to 1861 and as the female department of the Tuolumne County Hospital from 1866 to 1897.

In 1978, when this picture was taken, it was again a hotel, operated by Julia Ralston. It still serves as a hotel; as the Gunn House Motel, it has brought back the original name.

The Reverend Martin poses at the Sonora Methodist Episcopal Church in 1903. This church was torn down in the early 1920s.

The rustic red St. James Episcopal Church, built in 1860 and thought to be the oldest Episcopal Church in California, is located at the north end of the downtown area, where Washington Street divides.

Entrance to the Tuolumne County Historical Society Museum, originally the Sonora City Jail.

Many of the new buildings in Sonora are built so that they look old. Note the cast iron balcony railing on these buildings.

135

The grandstand at the Sonora Race Track, photographed in 1895. This facility remained in use until 1905.

The old oval track can still be seen in this 1978 view of the former race track property that is about to be developed.

Sonora Independent Hose Company #2, 1886.

The last of Sonora's old Chinatown. All of these old buildings were torn down in 1948. A monument made of bricks salvaged from one of the buildings was on this site, which is now a parking lot.

A 1914 view of the Standard Lumber Company mill and box factory at Sonora. This mill produced sash molding, lumber, boxes, shook, etc. The owner of this large concern was D. H. Steinmatz, and Joseph L. Dambacher was general manager. All of these buildings are gone now, and a Safeway store occupies part of the site.

A Stockton-built Holt Steamer was used by the Standard Lumber Company to haul logs and cut lumber after the turn of the century.

Interior of the hoist house at the Golden Gate Mine. At the left is a steam engine. The engineer stands next to a steam-operated hoist that operates the elevator in the shaft.

Superintendent Joe Frances and Louie Haferl (in the dark sweater) on one of the underground levels of the Golden Gate Mine.

The best known and most photographed building in Columbia, 1978. Records indicate that tons of gold, worth almost $87,000,000 at the rate of $20.67 per ounce, were weighed through this office. This was the first building restored when the town became a state park in 1945. The stagecoach excursion for tourists, complete with a bumpy ride and lurking badmen, is still exciting.

By 1860 the town of Columbia, three miles from Sonora, had forty saloons, four banks, three express offices, six breweries, eight hotels, two fire companies, three churches, two bookstores, three theaters (one Chinese), and a school. On July 15, 1945, in Columbia, Governor Earl Warren signed a bill creating the Columbia Historic State Park.

St. Andrew's Presbyterian Church (Columbia's "Church of the 49ers") was organized in 1854. It was built in 1864 and destroyed by fire on June 22, 1950. The building in this 1978 picture is an exact replica of the original and was dedicated in 1954, the centennial year of the church's founding.

This Columbia school cost $4,800 when it was built in 1860. It was used until 1937, and this view shows a class around the turn of the century. In 1960 the building was restored at a cost of $60,000. The California Teachers' Association and the schoolchildren of California raised $52,000 for the restoration.

The Tuolumne engine company and the Columbia Drug Store. The firehouse is the present home of the "Papeete," a hand-operated engine with a fire hose fashioned of riveted buffalo hide.

Saint Anne's Church as it looks today. This photo was taken in 1978.

Saint Anne's Catholic Church, built in 1856, overlooks the town from Kennebec Hill. The bell came from New York at a cost of $1,500. The photograph at left was taken before the church was restored in the 1920s.

The altar of Saint Anne's Catholic Church in Columbia was painted by Jim Fallon, a wealthy man and one-time owner of the Fallon Hotel and Theater.

Hydraulic mining near Columbia, probably at Knickerbocker Flat. Elaborate systems of dams, canals, reservoirs and pipelines channeled water under high pressure to gold deposits. The gravel and rock were washed into long sluices, where riffles held the gold. The jets of water directed onto the hillsides caused entire sections of cliffs to be undermined and collapse. Whole mountainsides were washed away, and the scars from this destructive mining can still be seen. Among the first important conservation laws in California was state legislation in the 1880s and 1890s putting an end to hydraulic mining.

The Administration Building at Columbia Junior College, located near Columbia State Park, is circular in design, with a balcony around the second floor giving a fine view of the lake in the center of the campus. The campus is about ninety percent complete, and in 1970 its architects received an award for its beauty from the American Association of Community Colleges. In 1979 Columbia Junior College had forty full-time faculty members, 2,500 students, and a curriculum of more than three hundred courses.

Standard, near Tuolumne, in 1914. The lumber mill burned to the ground in 1937.

Logging train hauling seven-foot to sixteen-foot fir logs to the West Side Lumber Company pond at Tuolumne about 1912.

The steam donkey engine was once standard equipment for taking logs up steep hillsides in the woods. The logs were hauled by oxen to a railroad siding on a traction engine. Mounted on a log sled, it moved the logs by means of a cable attached to the wench in front of the vertical boiler. The Westside Lumber Company of Tuolumne used donkey engines as late as 1930.

Carter's, a small mill town near Tuolumne, was established in 1901.

Tuolumne Square Shopping Center, on the outskirts of the city, in 1978. Many of the old Mother Lode towns are starting to build shopping centers like this.

Main Street, Tuolumne, in 1978.

Mark Twain's cabin on Jackass Hill was rebuilt in 1922 and is still standing. This area was a stopping place for pack trains enroute to the mines. The name Jackass Hill comes from the fact that as many as two hundred jackasses were tied up on the hill overnight at times.

James Gillis under the big oak tree on Jackass Hill.

Bill Gillis stands under this big oak tree to show its size. Jim and Bill Gillis were great friends of Mark Twain, who lived in their cabin on Jackass Hill during the winter of 1864-65, when he wrote his famous jumping frog story.

James Chaffee and Jason Chamberlain on the front porch of the so-called Bret Harte cabin in Second Garrotte. Their friendship may have served as the inspiration for Bret Harte's immortal story, "Tennessee's Partner." Chaffee died in 1912 at the age of eighty, and Chamberlain died two years later.

The Hangman's Tree in Second Garrotte, circa 1935. This settlement, first called Saint Ignacio, was located east of Groveland, on the edge of the Mother Lode region. The huge oak tree is said to have served as a gallows for sixty hangings.

Tuttletown was settled in 1848 by Mormon prospectors and was called Mormon Camp. A year later Judge Anson A. H. Tuttle built the first substantial house in the town, and the miners changed the name to Tuttletown in his honor. On the left is the Tuttletown Hotel, and on the right is the Swerer Store, built in 1852. This was the store where Mark Twain bought his supplies, as Tuttletown was near Jackass Hill, where Mark Twain lived with the Gillis brothers in 1865. This photograph was taken before 1940. Unfortunately, both of these buildings are now gone.

Confidence post office. A portion of the schoolhouse can be seen in the background. The last postmaster in Confidence was Junius B. Booth.

The surface works at the Confidence Mine, located between Twain Harte and Long Barn. Before it closed in about 1912, it had produced $4,200,000 in gold.

Chinese, employed by a group of Englishmen to work their claims, made up the bulk of the population in the town that became known as Chinese Camp.

The spot where gold was first discovered in Chinese Camp in 1849. A war between Sam-Yap (1,200 strong) and Yan-Woo (900 strong), fought here on September 26, 1856, was the second Chinese tong war in California. Four Chinese were killed in the war that lasted all day and was viewed by a large crowd.

Main street in Chinese Camp in 1978. To the right is the post office, built in 1855 and still in use today, serving about two hundred people. In its heyday it served a population of more than five thousand. Among the principal developments in the town were a race track, joss house, fandango house, saloon, dance hall, casino, opera house, Miner's Union Hall, Man-Chow temple, Garrett House, Wells Fargo Depot, Weir Brewery, Rosenblohm's Store, and many beautiful homes.

Garrett House was the terminus of a stage line owned by the Morris brothers. The line carried passengers from the railroad station in Chinese Camp to Yosemite Park and back. The Morris brothers auctioned off the equipment in 1903, and the railroad took over. This building is no longer standing.

The old one-room schoolhouse, left, built in 1906, and the new school, built in 1970 of Oriental design in tribute to the town's founders. Kindergarten through eighth grade are taught in the new school, which has two classrooms, one multi-purpose room, and two teachers.

A 1978 view of the old Saint Francis Xavier Church in Chinese Camp. It was built in 1855, and Father Henry Aleric was the priest. The church was restored in 1949. The cast iron fences protect gravesites, many of which bear dates in the late 1800s. The Chinese population did not follow the Roman Catholic faith and, although there were more than five thousand Chinese living in the area, the church had only seventy parishioners.

Knight's Ferry jail, built in the early 1850s, rests on bedrock and was made of iron throughout. Small openings in the doors were closed with iron flaps secured on the outside with long iron pins.

A 1920 view looking east along State Highway 120, toward Tuolumne County. The town of Knight's Ferry, on the north bank of the Stanislaus River, is on the extreme left.

Ruins of the Tulloch Grist Mill at Knight's Ferry, idle since 1899. The original building was washed away in the flood of 1862, and shortly thereafter was rebuilt of huge sandstone blocks. Today the mill is a hollow shell of its former self.

Knight's Ferry Methodist Episcopal Church, originally built in 1860, was rebuilt in 1908. This photograph was taken in 1978.

Main Street, Jacksonville, in 1915. Located about fifteen miles east of Sonora, along the south fork of the Tuolumne River, Jacksonville was first called Jackson's Camp. It was named in honor of Colonel Alden Apollo Moore Jackson, who had founded the town of Jackson in Amador County. In a short time Jacksonville became an important mining center, with its huge creaking waterwheels, flumes, rockers, long winding sluices, long toms, and other gold-extracting equipment. The few buildings remaining in the town were torn down before the area was flooded by the Jacksonville Reservoir.

One of the main power houses of the Hetch Hetchy system is located at Moccasin Creek near the junction of Highways 49 and 120. The combined waters from Hetch Hetchy Reservoir and Lake Eleanor plunge over 1,300 feet down the pipelines in the background to the penstocks.

Priest's Hotel was built atop long steep Priest's Grade, on the Big Oak Flat Road, during the late 1850s. It was a welcome resting place for people traveling by horse-drawn stages to and from Yosemite Park. The hotel was owned and operated by W. C. Priest. This picture was taken in the 1920s, just before the hotel burned to the ground.

The Eagle-Shawmut Mine, lcoated on the Tuolumne River along Highway 49, two miles north of Jacksonville, in 1898. These buildings were long gone before this area was covered by water upon the completion of the new Don Pedro Dam. One of the larger mines, the Eagle-Shawmut produced over $9,000,000 in gold from 1897 until it closed in 1942.

The Shawmut Mine Chlorine Works, located along the old Shawmut Grade near Chinese Camp. The foundations of these buildings are now under the waters of the Jacksonville Reservoir.

VII

Rivers of Tuolumne

The rivers of Tuolumne County have played a vital role in its history; in fact, the rivers are where the county's modern history began. Tuolumne County has some seven hundred miles of streams, but the Tuolumne and Stanislaus Rivers are the two great waterways that have been so important to the county's development in many ways—gold production, recreation, water power and irrigation.

Gold is what made it all happen, of course, and it was in a river that the first gold in California was discovered. The gold was formed in prehistoric times, and the old rivers first loosened the gold from its rock crevices and mixed it with the sands and gravel of their beds. With the next upheaval of the earth, when the Sierra Nevada Mountains were formed, the newer and more rapid waters of the rivers and creeks of the Mother Lode increased the amount of free or placer gold.

The Tuolumne River rises in the glaciers and high peaks of the Sierra Nevada, in Yosemite National Park. Much gold was found along its gravel bars during the early placer mining days of the gold rush, and settlements were established at nearly every one. The most important of these sand bar camps was probably Jacksonville, which lasted until it was flooded by the waters of the Don Pedro Dam.

Even after the gold supply was exhausted and mining activity ceased, the Tuolumne River remained attractive as a source of irrigation for neighboring farmlands. The Turlock Irrigation District and the Modesto Irrigation District joined together in the 1880s to build a dam at La Grange, and the first waters were diverted in 1900. The two districts constructed the first Don Pedro Dam and power plant in 1923, and then Turlock built a power plant at La Grange.

Even after the La Grange dam had been built, the Tuolumne River had more water than was being used, and this additional supply was sought by the City and County of San Francisco, which had lost its previous water source in the tragic earthquake and fire of 1906. San Francisco obtained permission from the United States Government to store Tuolumne River water in the Hetch Hetchy Valley in the Sierra Nevada and transport the water under the bay and into San Francisco by aqueduct. Diversion of the Tuolumne waters began in 1934.

But there still was water from the river remaining uncollected, and it was needed. In 1967 the two irrigation districts and San Francisco launched the second Don Pedro project. This new and much larger dam, of earth-rock fill, inundated the old dam and powerhouse, located one and one-half miles upstream. The new dam, the sixth largest in California, is 580 feet high, with a crest of 1,900 feet containing 16,000,000 cubic yards of material. The new power plant contains three 45,500-kilowatt generating units driven by three 70,000-horsepower turbines.

Additional power is only one benefit provided by this vast project; another is additional recreational facilities. The lake behind the dam is 26 miles long and has a shoreline of 160 miles. It covers nearly 13,000 surface acres and provides ample opportunities for boating, fishing, camping and picnicking. It is also an additional source of irrigation water. Thus, the river still produces "gold"; not the yellow gold that sparked its development, but the dollars of the tourists, farmers, and cities who use its waters for recreation, irrigation, and power.

The three-branched Stanislaus is the other historic river of Tuolumne County. It forms the boundaries between Tuolumne County and Stanislaus and Calaveras Counties.

According to legend, the Stanislaus is named for the great Miwok Indian chief, Estanislao. He had been a neophyte with the padres at the Mission San Jose, where he had been given his Christian name, but he ran away from the mission and, with the aid of so-called "wild" Indians, began a series of very profitable raids on the missions. Estanislao was a great leader until his defeat by General Vallejo's troops in a battle at the junction of the Stanislaus and San Joaquin Rivers. Thanks to the intervention of Father Duran of the Mission San Jose, the Mexican governor pardoned Estanislao on the condition that he steal no more mission stock.

Like the Tuolumne River, the Stanislaus owes its early development to gold mining. This river is in the heart of the

Southern Mines, and literally dozens of gold camps sprang up along its sand bars. In time, some of these camps became important ferry crossings and bridges. It is said that Mead and Robinson built a ferry crossing at what was to become Melones and took in over $10,000 in ferry fees in the year 1848. Later, Percy Wood acquired the ferry and operated it until Calaveras and Tuolumne Counties purchased it and built the first bridge on what was to become the Highway 49 crossing.

Major crossings and towns sprang from other ferries on the Stanislaus River. Knight's Ferry, an important mining community in the 1850s and 1860s, still attracts tourists. Of special interest there is the Old Covered Bridge, built after the great flood of 1862 washed out all the bridges on the Stanislaus.

Another crossing, a few miles above, on the road between Stockton and Sonora, was operated by P. O. Byrnes and became known as O'Byrnes Ferry. In the 1860s Byrnes built another covered bridge that eventually became a public bridge. It lasted until it was finally torn down in the 1970s as the area was flooded by Tulloch Reservoir, a part of the $52,000,000 Tri-Dam project. This dam was dedicated on June 15, 1957, and was for the benefit of the Oakdale and South San Joaquin Irrigation Districts.

O'Byrnes Ferry is said to be the locale of Bret Harte's "Outcasts of Poker Flat." The site of his other famous short story, "The Luck of Roaring Camp," is supposedly a few miles upstream. In fact, Harte is supposed to have situated more of his stories on the Stanislaus than on any other river.

Parrotts Ferry is another historic Stanislaus River crossing. The bridge there, on the road from Vallecito to Columbia, has been important since the 1850s, when these towns were major mining centers.

Today the crossings at both Parrots Ferry and Melones are beautiful new high-level bridges to make way for the great New Melones Reservoir. The Archie Stevenot Bridge at Melones has been called the most beautiful bridge in all the Mother Lode, and certainly on historic Highway 49. It cost $15,000,000 and was named for "Mr. Mother Lode," Archie Stevenot, who was born at Carson Hill. He lived all his life in Calaveras and Tuolumne Counties, and was the one person most responsible for the designation of Highway 49 as the Golden Chain Highway. He organized the Golden Chain Council to promote a better link among the gold counties. He served as president of the Council and was on the board of directors for forty-three years. The State Legislature named him "Mr. Mother Lode" for this important work.

These new high-level bridges are necessary because the 625-foot earthfill New Melones Dam will provide a 2,400,000 acre-foot reservoir. The Army Corps of Engineers built the dam, which will be operated by the Federal Bureau of Reclamation. The total cost of the project, nearing completion as this book goes to press, will amount to $340,000,000, including road, bridge and power development. The high-level bridge at Parrots Ferry cost $10,200,000, and the Camp Nine road and bridge a few miles farther up the river cost $2,800,000.

The dam's 300-megawatt hydroelectric plant is capable of generating more than 430,000,000 kilowatt hours of energy per year, enough to supply the electrical needs for a city of 200,000 people.

The reservoir began to fill in the fall of 1978, but filling was restricted to accommodate the archaeological mitigation project conducted at the site by a private firm under contract from the United States Government. The project has proceeded from the bottom up, so the lower elevations of the reservoir basin could be filled as the project finished at these levels. The program was funded by recent legislation that authorizes the spending of one percent of the total cost of a project for archaeological, cultural, or historical mitigation.

The reservoir collects water from a 900-square-mile area and is designed to completely cover the old Melones Dam. Although the first season's runoff was not enough to fill the reservoir, it was enough to submerge the old dam. With the drop in the water level resulting from the 1979 dry season, however, parts of the old dam were exposed, but it is anticipated that subsequent runoffs will fill the reservoir and keep the water level high enough to keep the old dam covered.

The new dam has generated considerable controversy. Some people have questioned its necessity, and strong protest has come from "wild rivers" supporters, as the area from Camp Nine, near Vallecito, to Melones has become very popular with rafters who ride the river rapids in rubber boats. If the reservoir fills as projected, this particular stretch of river will be under many feet of water, but whether or not this will happen will be determined by the results of litigation now pending in the courts.

The reservoir, however, with its 100 miles of shoreline, will produce a great new recreation area of approximately twenty square miles. Camping, boating, fishing, swimming, and water-skiing facilities will attract many visitors. The tourist dollars will be "new gold" for the residents of Tuolumne and Calaveras Counties, who will also enjoy these new recreational opportunities themselves.

The most modern touch on the one-way Knight's Ferry covered bridge is an automatic signal light system installed in the late 1960s. The bridge is covered to protect the all-wood trusses from the weather and thus greatly lengthen the life of the bridge.

This 1978 photo shows one of California's old covered bridges stretching across the Stanislaus River at Knight's Ferry. This bridge was built in 1862, the same year a flood washed away the original bridge, built in 1854. This replacement bridge is still open to traffic, but there are plans for a new bridge upstream from it, and when that one is completed, this old bridge will remain as an important part of a historic park now being planned for this portion of the river.

The old bridge was sold at auction in 1957. Plans to preserve it were ruined when the wooden span broke while it was being floated down the river. Some of its timbers were used by Cliff Mitchel in building the Black Creek Lodge, one of the first resorts at Tulloch Reservoir, in 1957-58. This facility is now operated as the Copper Cove Resort.

Taken in 1957, this photo shows the 1862 O'Byrnes Ferry covered bridge on the old Mountain Pass Road from Copperopolis to Jamestown.

A close-up view of the second O'Byrnes Ferry, a covered toll bridge built in 1862, the same year a flood washed out the first one. In 1902 Calaveras and Tuolumne Counties purchased this span and made it toll-free. It remained in use until 1957, when it was replaced by a new high-level bridge that is part of the Tri-Dam project.

At the far left is the O'Byrnes Ferry covered bridge on the Stanislaus River, just before it was removed to make way for the Tulloch Reservoir in November 1957. The new bridge in the foreground is 3,000 feet upstream. All of this valley is now covered by the Tulloch Reservoir.

Tulloch Reservoir was named for Charles Tulloch, a pioneer rancher. It is part of the Tri-Dam project of the Oakdale and South San Joaquin Irrigation Districts. The new concrete bridge is 620 feet long and 20 feet wide.

These ruins at Poker Flat are located at the west end of the O'Byrnes Ferry covered bridge on the Stanislaus River. It is claimed that Bret Harte used this setting for his famous story, "Outcasts of Poker Flat." In the background is the lava flow he called Table Mountain.

One of the first ferry crossings of the Stanislaus River at Melones. The ferry was operated by John Robinson and Stephen Mead. Later, in the summer of 1849, they also established a trading post.

The original Don Pedro Dam was built on the Tuolumne River, just below Highway 49, in 1923. This 1,040-foot-long dam and its powerhouse were inundated in 1970 when the new Don Pedro Dam was completed one and one-half miles downstream. The latter project is a joint venture of the Modesto and Turlock Irrigation Districts and the City and County of San Francisco.

Archie Stevenot was born in 1882 near Carson Hill. He served as general superintendent of the Carson Hill Gold Mine and was also a rancher, postmaster, and miner. He died in 1968 in Sonora.

A 1978 air view of the New Melones Dam and Reservoir, located on the Stanislaus River about one mile below the old dam. The first Melones Dam was built in 1926 by the Oakdale and South San Joaquin Irrigation Districts to provide irrigation water and electric power. The new concrete structure was started in 1966 and completed on October 28, 1978.

This aerial view shows the Archie Stevenot Bridge on Highway 49 at Melones. The reservoir now covers the older bridge at the lower right. The new 440-foot bridge, which opened on November 22, 1976, has been widely praised for its beauty.

VIII

Mariposa

Mariposa apparently received its name from the large number of beautiful multicolored butterflies seen by the famous Spanish explorer, Gabriel Moraga, on his trip into the great Central Valley in 1806.

Mariposa's fame stems from many sources. It is the southern terminus of the fabulous quartz vein called the Mother Lode and has a rich history of gold production. The Fremont Grant, or Mariposa Grant, of ten square leagues (45,000 acres) covered most of the great mines in Mariposa County.

Another reason this region is famous is that when the first twenty-seven counties were created by the Legislature in February 1850, a vast amount of land (over thirty thousand square miles) was designated as Mariposa County. As time passed and California's population increased, parts of nine other counties were eventually formed within its boundaries. Historians have called Mariposa the "Mother of Counties."

When the first county boundary lines were drawn, Mariposa spread from the summit of the coast range to the present state of Nevada, and from Tuolumne County on the north to the summit of the Tehachapi Mountains on the south. It included Mt. Whitney and Death Valley, the highest and lowest spots in the country. In 1850 the county seat was first located at the little mining town of Aqua Fria, four miles west of Mariposa, but as the miners' population shifted, the county seat was moved to the settlement of Mariposa. The original spelling of Aqua Fria was the Spanish Agua Frio.

A stately classical courthouse was built in 1854 for $9,000. Although the belltower was built in 1854, the present clock was not installed until 1866. It has been used ever since. This remarkable building, the oldest courthouse in continuous use in California, still contains much of the original courtroom furniture.

Many famous cases were heard here, including Biddle Boggs vs. Merced Mining Company, the leading case expounding the doctrine that the precious metals belonged to the State by virtue of her sovereignty.

A weekly newspaper, the *Mariposa Chronicle*, was first published in the same year that the courthouse was built. The name was soon changed to the *Mariposa Gazette*, and this 125-year-old weekly newspaper is still telling the news in the land "above the fog, below the snow."

In 1918 John L. Dexter acquired the *Gazette*. He and his family have chronicled the news in this outstanding weekly since then. In 1948 his daughter, Mrs. Margaret Campbell, took over, and her son, Dexter Campbell, has been the publisher since 1974. The original *Gazette* building houses an exhibit at the County Fairgrounds.

In July 1858, the *Gazette* recorded the exciting conflict between the forces of John C. Fremont of the Mariposa Grant and the army of the Merced Mining Company. They were fighting over the rich Bear Valley mines known as the Josephine and Pine Tree. After losing the Josephine, Fremont was able to hold possession of the Pine Tree physically until the Supreme Court ruled in his favor.

Colonel John C. Fremont, "the pathfinder of the West," was an important citizen of Mariposa County in the exciting gold mining period of the 1850s and 1860s. While Fremont was participating in the conquest of California in 1846 and 1847, and before he was court-martialed by General Stephen Kearny, he gave his friend Thomas O. Larkin $3,000 to purchase a Mexican land grant near San Francisco Bay. Instead, Larkin purchased a "floating" grant of 45,000 acres from Juan Alvarado, who had received it from Governor Micheltorena. It was situated in the dry eastern foothills of the interior valley and was known as "La Mariposa."

Although Fremont was indignant at Larkin's action, he was stuck with the grant but made no effort to locate its boundaries until after he learned that rich gold strikes had been made in the vicinity of his grant. He succeeded in locating the boundaries of the grant so that it lay across the famous quartz vein of the Mother Lode and included several rich mines. Because of its peculiar shape, it is sometimes called the "Frying Pan Grant." Miners who had prospected and developed that land contested Fremont's claim to it. After several years of legal contest and even violent conflict, Fremont's grant was confirmed by the American court in 1859.

The richest mines on the grant were the Mariposa, the Pine Tree and the Josephine, and also the Princeton at Mt. Bullion. The Mariposa produced about $3,000,000 in gold, and the Pine Tree and Josephine, operated together, produced over $4,000,000. It has been estimated that at least $15,000,000 in gold was produced from mines on this grant and that there was a total gold production of $48,000,000 for all of Mariposa County. In 1863 Colonel Fremont sold the grant to a New York banker. The purchase price was claimed to be $1,500,000.

An important part of Mariposa County history is the discovery of Yosemite Valley. Yosemite was probably first glimpsed by the Joseph Walker party in 1833 and then seen from the west end by gold miners in October 1849. The Mariposa Battalion, however, was the first group to enter, explore and publicize the wonders of the valley. They found Yosemite while chasing Chief Tenaya, leader of the Yosemite Miwok Indians.

The discovery was an indirect result of the friction that developed as settlers and gold miners came into the region. After several miners were killed, Governor John McDougal ordered Sheriff James Burney to recruit two hundred men, capture the Indians, and put them on a reservation near what is now Fresno.

In 1851 the Sheriff passed the assignment to Major James Savage, who had been attacked by the Indians at his trading post on the Merced River. Savage recruited men and formed the Mariposa Volunteers or Battalion to bring the Indians into the reservation. Some refused to come in, and on March 25, 1851, while trailing the Indians, the Battalion first saw the "Incomparable Valley" from a spot near old Inspiration Point. Dr. L. H. Bunnell, a member of the Battalion, later described his sensation upon seeing the valley for the first time: "As I looked, a peculiar exalted sensation seemed to fill my whole being and I found my eyes in tears of emotion."

As a result of this expedition, news of Yosemite's beauty spread and a worldwide interest in the valley developed. Mariposa, as the gateway to Yosemite National Park, became important as a tourist center, supplying thousands of visitors to the Park with food, fuel and lodging.

Two famous scenic highways intersect at Mariposa, State Highway 140, from Merced to Yosemite Valley, and State Highway 49, traversing the Mother Lode Country from Oakhurst on the south to Vinton in Plumas County on the north.

One of the most significant recent developments in Mariposa County was the formation of the Mariposa County Historical Society in 1957, as an aftermath of the Mariposa Centennial in 1954. Judge Thomas Coakley was not only active in the formation of the Society but was its first president and largely responsible for its success. In 1959 the Society set up an interesting museum in the rented Masonic Hall. It attracted many visitors, but facilities were very limited.

Hopes for a larger building got a tremendous boost in 1967, when two members of the Society, Merle and Clay Daulton, gave $25,000 for a new building. The Society set out to raise a total of $125,000 to construct a fire-resistant building to house a museum and a public library. The County Supervisors matched the $25,000 gift, and the Mariposa County Historical Society raised the additional $75,000 from friends and well-wishers in two years. Many people and firms contributed time, services and materials, and on May 23, 1971, the dream was realized with the dedication of the new Mariposa History Center.

Today, the History Center recounts the history of Mariposa County through outdoor exhibits of an Indian sweathouse, a five-stamp gold mill, early buildings, and an artistic array of exhibits and artifacts in the main building. No one interested in the history of Mariposa County should miss the museum.

Outlying areas such as Hornitos, Bear Valley, Mt. Bullion, Mt. Ophir, Bagby and Coulterville are also part of Mariposa County history.

According to prominent Mariposa County historian Shirley Sargent in her *Mariposa County Guidebook*, Hornitos was created when Mexican gamblers, miners and dance hall girls were forcibly evicted from nearby Quartzburg in 1850. Thomas Thorn had found gold in Quartzburg in 1849 and gave the town its name because of a nearby quartz outcropping. However, it was Hornitos that survived, as placer gold was discovered on Burns Creek, which passed through the settlement. The town was named Hornitos because of the peculiar graves in the cemetery, which were mostly above ground due to the rocky soil. These cairns resembled little ovens. Several of them still survive.

Hornitos was one of the toughest towns in the Southern Mines and was supposedly a hangout for the legendary Joaquin Murietta and his bandits. A tunnel he is said to have used to escape from a fandango house is still visible and marked.

Later Hornitos gained respectability and became a stage stop and trading center. Domingo Ghirardelli, later to become known as the "Chocolate King," had a store there, and its ruined walls are also marked.

The town of Bear Valley was significant in the early history of Mariposa County, as John Fremont chose it as the headquarters for activities on his Mariposa Grant. He brought his wife Jessie and three children here to live in the "Little White House" in 1858-1859. Fremont chose the name for the Oso House Hotel. *Oso* is Spanish for bear.

Mt. Bullion was called Princeton at one time, for the important quartz mine nearby. The Princeton was one of the biggest gold producers and employers in the county. In 1862 the post office and town were named Mt. Bullion, for Senator Thomas Hart Benton, Fremont's father-in-law, whose advocacy of "hard money" earned him the nickname "Old Bullion." When the mine was running Mt. Bullion was a busy and thriving town, but today only Trabucco's Store, the Princeton Saloon and the Mt. Bullion School remain, and they are all closed.

Nearby are the remains of the Mt. Ophir private mint of Moffat and Company. In 1851, due to the shortage of coin, it was authorized to strike off fifty-dollar gold pieces.

A year earlier, George W. Coulter and George Maxwell had begun operating stores in a Mexican mining settlement not far away. Because Coulter displayed two small American flags in the tree near his trading post, the Mexicans called the town "Bandarita." Coulter and Maxwell later drew straws to determine whose name the town should bear. Coulter won, and the town was named for him, but Maxwell's name was given to the creek and the post office.

In time Coulterville became a thriving community of five thousand, including large Mexican and Chinese populations. The area is said to have produced about $15,000,000 in gold. Reliable sources state that the town supported ten hotels and at least twenty-five saloons when it was busy with miners, freighters and tourists to Yosemite Valley.

Several disastrous fires reduced the town's size, but it is still an active tourist center. The Jeffery Hotel is the most important building in the town. President Roosevelt registered here in 1902. The Magnolia Saloon next door boasts collections of guns, minerals and other artifacts. Ed and Vi Sackett, who took over the Magnolia in 1944, are responsible for the collection.

Across the street, in the Wells Fargo Building, is an interesting museum run for years by old Vernon Pepper. It is now under new management. In front of the building is the hangman's tree that shades "Whistling Billy," the eight-ton mine locomotive that served the Mary Harrison Mine. This mine opened in 1853 and operated to the 1,200-foot level. It was one of the large gold producers in the county, with over $1,500,000 to its credit. The ore averaged $7 to $12 per ton when gold was $20.67 per ounce.

The beams in the framework of the 1854 Mariposa County Courthouse are fastened together with wooden pegs. All of the lumber used in this building was produced locally.

The courtroom of the Mariposa County Courthouse has been in continuous use since 1854. The judge's bench is nine feet long to accommodate the three judges of the old Court of Sessions. The bench has many dents from the judges' gavels.

The Mariposa Courthouse as it looked in 1884, before the cedars grew. The people posing for this picture are county officials. Judge Crocker, the first judge in Mariposa County, is fifth from the left, in the top hat.

The Victorian home of Superior Court Judge J. J. Trabucco still stands today, just a block from Main Street.

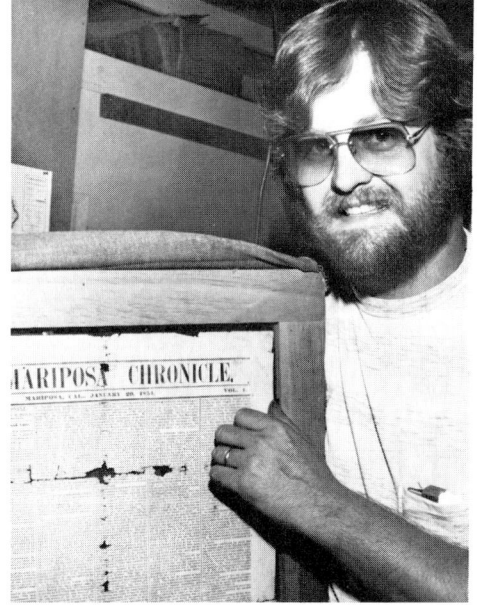

Dexter Campbell, present publisher of the Mariposa Gazette, *displays part of an 1854* Mariposa Chronicle. *One year later the name was changed to the* Mariposa Gazette.

John Dexter, owner and publisher of the Mariposa Gazette, *on the front porch of the newspaper office. Mr. Dexter, born near Coulterville, was superintendent of schools in Mariposa County before he bought the* Gazette *in 1918. The old* Gazette *building has been moved to the Mariposa County Fairgrounds.*

Main Street in Mariposa on October 18, 1930, during the re-enactment of the "Days of '49." The two-story building in the center formerly housed the Mariposa Museum.

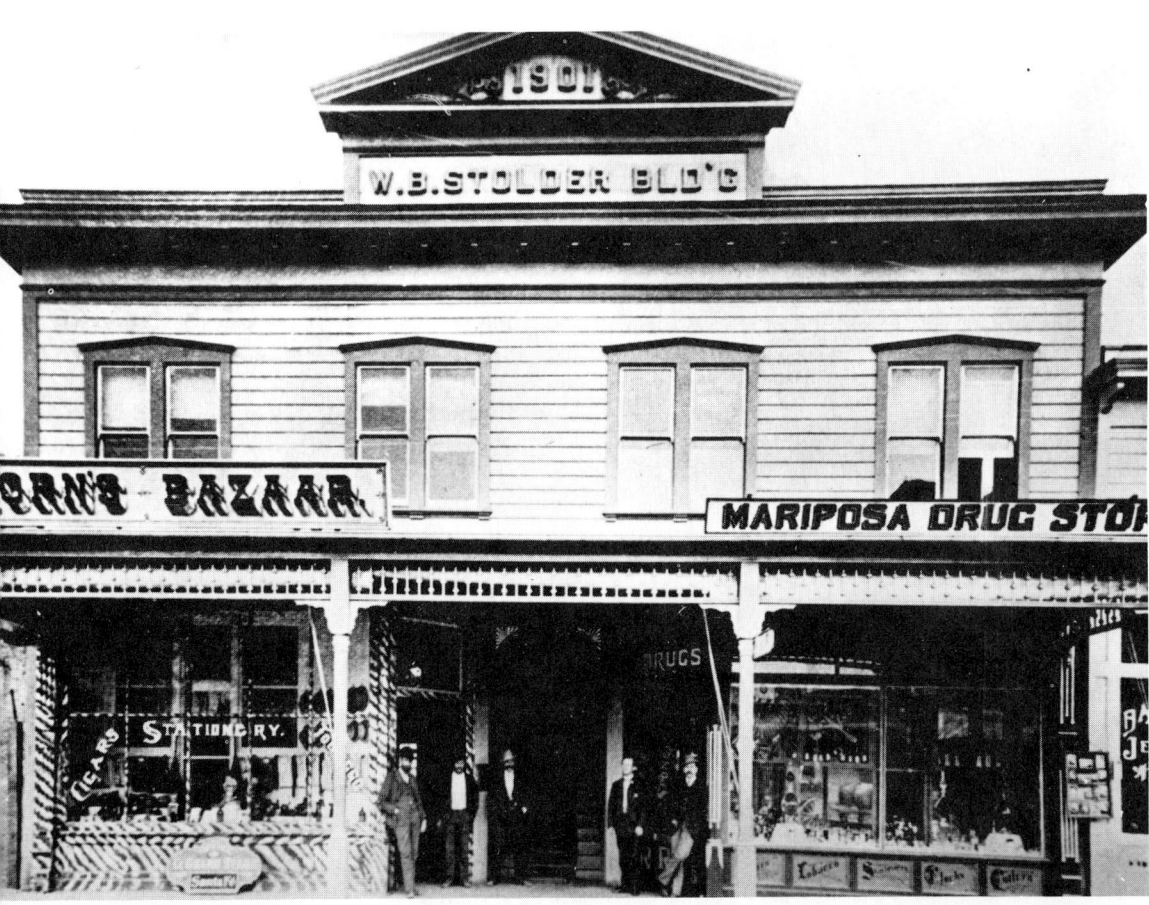

One of the more prominent commercial buildings on Mariposa's Main Street, as photographed about seventy-five years ago. This building later housed the town's first museum.

Mariposa's St. Joseph's Catholic Church was built in 1862 under the supervision of Father Auger. It has been in continuous use since it was dedicated January 18, 1863, by Archbishop Alemany.

The Mariposa County Jail was built in 1858 during the days of gold mining and outlaws, when counties executed their own criminals. The jail is made of native stone and still stands today, one block west of Main Street.

This structure was built in Mariposa in 1883 as a Methodist church. In 1966 it was moved to its present location for use as a Christian Science church.

It was not until the early 1850s that miners started to bring their families to the Mother Lode. They were full of adventure and intended to make a fortune. Some did, of course, but most didn't.

The first water-powered quartz mill built in California, on the Mariposa grant. In his book The Call of Gold, *Newell D. Chamberlain writes that the mill, transported across the Isthmus in 1850 for Commodore Robert F. Stockton, was located one mile from Mariposa on Stockton Creek. The city of Stockton was named after Commodore Stockton.*

General John C. Fremont, "the Pathfinder." When he returned to his Bear Valley land grant in 1857, after his unsuccessful campaign for the Presidency in 1856, he found many squatters settled on his land. Because the deposits were shallow and operating costs were high, the mines operated on his grant in later years were not always financially successful. Fremont died penniless in New York in 1890, but his name has been given to many parks, schools, and streets, and to the town of Fremont, California.

Jesse Fremont, wife of John C. Fremont, became famous for her letter-writing ability and added to the family's fame.

A map of Mariposa County, with the famous 45,000-acre Fremont Grant shown in the shaded area.

Galen Clark, seated in the center of the wagon going through "Wawona" in the Mariposa Grove of Big Trees, the most famous tree in the world. This tree is no longer standing; it was the victim of a severe windstorm a few years ago.

Driving through the "Wawona Tree."

Galen Clark, "Guardian of the Valley," on August 12, 1909 at the age of ninety-five. In 1857 Clark discovered the Mariposa Grove of Big Trees and helped to make it famous. Ill and feeling he would not live much longer, he selected his gravesite and quarried his own tombstone. Fifty-three years later he was buried there as he wished, with other pioneers in the little graveyard near Yosemite Falls.

*Yours Sincerely
Galen Clark
Yosemite Calif.*

Galen Clark's signature in his book, Indians of the Yosemite.

The famous Mirror Lake at the upper end of the Valley, where sunrise services are held every Easter. Automobile travel to this point is now barred; visitors can reach the lake only by shuttle bus.

John Muir, the famous naturalist and conservationist, promoted interest in the beauty and wonder of Yosemite and the adjacent High Sierra. He was sometimes called the "Father of the National Park Service" because he advocated government protection of the scenic wonders of the Sierra Nevada Mountains.

The Army had control of Yosemite until 1864, when this beautiful valley became a state park. It became a national park in 1890.

A view of Yosemite Valley from the Wawona Tunnel. Near here is old Inspiration Point, where Jim Savage and the Mariposa Battalion first saw the Valley in March 1851.

Yosemite National Park, in the heart of the Sierra Nevada, has an area of 1,189 square miles. Every year more than one million people, from every part of the world, spend their vacations here. Pictured are the famous Yosemite Falls.

The new $125,000 Mariposa History Center is at the north end of town. Construction of this important facility was financed largely by donations collected by the Mariposa County Historical Society.

A five-stamp gold mill on the grounds of the Mariposa History Center. The mill was originally used in this area, and it is still put into operation on special occasions.

A 1928 view of the plaza in Hornitos. Some of these buildings still stand today. At one time Hornitos had a population of fifteen thousand.

Ruins of the Domingo Ghirardelli store in Hornitos, shown in the 1940s. It was here that the San Francisco chocolate king began his career.

The Hornitos Hotel, shown in 1862, was owned by E. G. Wells, the Wells Fargo agent. Wells was a popular innkeeper, and his hotel became a favorite stopping place for many notables touring the Southern Mines, including General Ulysses S. Grant and Vice President Schuyler Colfax. In the 1930s the hotel was razed for its lumber.

An artist's sketch of the legendary Joaquin Murietta who, according to old-timers, visited the fandango and gambling houses of Hornitos.

The old Mexican cemetery on a hilltop at the edge of town. From a distance the tombs look like ovens; hence the name Hornitos, Spanish for "little ovens."

A 1978 view of the old cemetery.

St. Catherine's Catholic Church stands on a hill overlooking Hornitos. This clapboard structure is shown here in a 1978 photograph. Buttresses were added recently to strengthen the walls.

A 1978 picture of the Hornitos Jail, a one-cell affair about twelve feet square. It had a dirt floor and no bunks or chairs. Prisoners were held for one day and night only, and then transported to Mariposa for trial.

The Oso House in Bear Valley, eleven miles northwest of Mariposa, was headquarters for John Fremont's famous guests and for his mining operations. The house was destroyed by fire in 1938.

The Fremont home in Bear Valley, where John Fremont and his wife and three children lived for several years after 1858. Local residents called the whitewashed cottage the "Little White House." It burned in 1866.

The Bear Valley jail was built of schist slabs in 1850. The floor was packed earth.

Bagby, formerly Benton Mills, was established on the Merced River in 1857 and had the first water-powered ore mill in California. Fremont's water-power stamp mill was named for his father-in-law, Thomas Hart Benton.

The bridge over the Merced River at Bagby, the route to Benton Mills. This location is now at the bottom of Lake McClure Reservoir on the Merced River.

Benton Mills Fremont Tramway (lower left) in the Fremont Grant, between Bear Valley and the Merced River.

The hoist works and mill at the now-vanished Princeton Mine at Mt. Bullion, 1900. First worked in 1852, the Princeton produced more than $4,000,000 in gold. It was not only the most productive mine on the Fremont Grant but also was one of the richest in the entire area. The Princeton mine was located just south of the present townsite.

A bird's-eye view of Mt. Bullion in the late 1800s. The town was called La Meneta when it was first settled in 1850, and later it was named Princeton.

In 1851 John L. Moffatt received official authorization from the United States Government to operate the first private mint in California. The mint produced hexagonal fifty-dollar gold slugs as a more convenient means of exchange than gold dust.

The Mount Ophir Mill and Mint was responsible for the development of a thriving town in the 1850s. Only ruins and a building or two mark the site today.

A birds-eye view of Coulterville. The Jeffery Hotel is left of center, and the old Coulterville Hotel is to the far right. The old highway can be seen starting on the left, running through the town, and exiting to the right. This picture was probably taken around the turn of the century. Note that there is not one auto to be seen.

Coulterville's Jeffery Hotel in 1978. Some of its walls are of rock and adobe thirty inches thick. First built in the early 1850s as a Mexican hotel, the building was altered into its present form by George Coulter in 1870.

Coulterville in 1978, seen from the improved Highway 49. Mexicans mined in this vicinity as early as 1849.

Mr. and Mrs. George Coulter, founders of Coulterville. Their home still stands on the outskirts of town.

An early view of Coulterville, first called Bandereta (Little Flag), looking south towards Mariposa.

Bruschi Bros.' general merchandise store was started by Francisco Bruschi and his wife Rose in the early 1850s. The miners welcomed Bruschi's silver-belled pack train, for it meant grub and beer were on the way.

The Coulterville I.O.O.F. Hall in 1978. This lodge was strong in the Mother Lode mining towns.

Evidence of the large Chinese population is the Sun Sun Wo Company store, located on the north edge of Coulterville. Built in 1851, it still stands today.

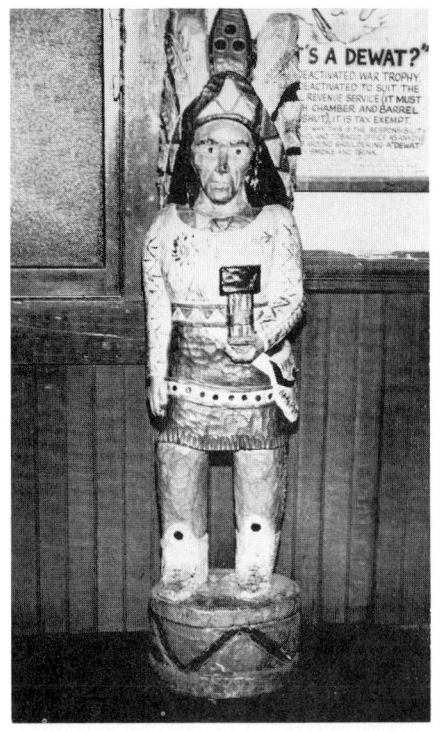

One of the items of interest in the old Magnolia Saloon Museum in Coulterville is this reproduction of a hand-carved wooden Indian, put in front of a store to advertise cigars.

Ore specimens are on display in the Magnolia Saloon, next door to the Jeffery Hotel in Coulterville.

The present Northern Mariposa County History Center was formerly Pep's Museum in the old Wells Fargo Building.

The "Whistling Billy" steam locomotive is hauled to the Mary Harrison gold mine, four miles north of Coulterville, by a twelve-mule team.

"Whistling Billy." This eight ton locomotive hauled fifteen cars, each loaded with five tons of ore. During its term of service it carried over $1,500,000 in gold ore. It came by boat to Stockton, then by rail to Waterford, and by logging wagon from there to Coulterville.

Horseless carriages lined up for a picture in Coulterville before starting on their way to Yosemite Valley in 1913. The first usable road to Yosemite was completed in 1874, through Coulterville. The first automobile to enter Yosemite Valley was Oliver Lippincott's Locomobile on June 24, 1900.

BANDS, ENTERTAINMENT AND SPECIAL EVENTS
In The Mother Lode Country

Stockton and Sacramento are the gateways to the Southern Mines, and people from all over the world passed through these two towns on their way to the gold country. Many of these argonauts were accomplished musicians, and it wasn't long before brass bands were formed throughout the Mother Lode.

One of the most popular bands was the Murphys Independent Cornet Band. They were in demand constantly, playing in many of the gold camps, and they also traveled to Stockton and Sacramento to play for concerts, parades and political rallies.

Besides the brass bands, many theatrical groups toured the gold camps, playing in saloons, gambling houses and makeshift theaters. It was amazing what large crowds attended these presentations. Imagine rough-and-tough miners sitting through a Shakespearean reading, understanding and enjoying it!

With the failing of the mines in the 1860s, many of the musicians returned to the country towns to engage in agriculture, boating and boat building. They still maintained their musical interests, and a number of their descendants continued with music.

As the roads improved, Stockton orchestras started to tour the Mother Lode country, playing for dances. In the late 1920s and 1930s orchestras such as Polly Watson, Leo Liberty, Charley Raggio, Patten and Springer, Art Caviglia and others played regularly, on Saturday nights and holidays, in the central Mother Lode.

The Utica Mine Brass Band of Angels Camp performed in front of the I.O.O.F. Hall in Sonora for the Admission Day Parade of 1895.

The Coulterville Concert Brass Band, sixteen strong, performed at all special events around Mariposa County.

Copperopolis Drum Corps, 1869.

One of the original hats worn by members of the Murphys Cornet Band.

The Jackson Military Band posed for a picture in front of the State Capitol in Sacramento in 1912. Archie Moore was the trumpet player and director.

The popular Murphys Cornet Band in concert at the Big Trees Hotel in the 1880s. The band is pictured atop the "Father of the Forest," the largest fallen tree in the Big Trees Grove. On the ladder in front of the band is the famous John Muir, who was visiting the Grove at the time.

The Fallon House, built by Owen Fallon, was one of the best hotels in the Mother Lode. Since 1949 it has been used by the University of the Pacific Drama Department and has become famous throughout the Mother Lode. Several of today's theatrical stars learned some of their skills in this old building.

The old Ratto Theater on Main Street in Jackson in the early 1920s. Note the poster on the left about everybody's favorite at that time, Tom Mix. Although many of the Mother Lode towns have many forms of entertainment—shopping centers, flea markets, rock concerts, the Jumping Frog Jubilee, gourmet restaurants, etc.—Jackson is one of the few that has been without a movie theater for years.

Lola Montez was a famous actress who entertained in most of the Mother Lode area during the Gold Rush period. Her home in Grass Valley is still standing and is preserved as a historical landmark.

Edwin Booth, a great Shakespearean actor, performed in many of the gold camps. He was a brother of John Wilkes Booth, who assassinated President Lincoln.

In her youth Lotta Crabtree played in the Angels Camp Theater and throughout the mining regions. She went on to become a famous actress.

The Stickle's Theater, Circus Hill, Angels Camp, decorated for a church fair in 1887.

The famous water tower featured in the long-running television series, "Petticoat Junction," is still standing today a few yards south of the Jamestown Depot.

The old Melones Bridge on the Stanislaus River was used many times as a set for Hollywood movies. The one above was used in a movie starring Will Rogers. This site is now under the waters of the New Melones Reservoir.

The old Drytown Town Hall in Amador County is now the Piper Playhouse, home of the Claypipers, a Bay Area company that performs old-time melodrama on summer weekends. This 1979 view is framed with the old firebell in the foreground.

The three-piece Patten and Springer Orchestra, playing for a dance in Valley Springs in 1899. Eli Springer is playing the violin and Bob Patten the piano. The trumpet player is unknown. Patten and Springer was the first band from the Stockton-Lodi area to perform throughout the Mother Lode. They traveled by horse and buggy, playing from town to town, and they drew the largest dance crowds of any orchestra.

The Art Caviglia Night Hawks played regularly every Saturday night in Love Hall on Main Street in Jackson. The dances went from 9 p.m. to 3 a.m. (or later!). During the dinner hour, from midnight to 1:00 a.m., everyone would meet at Buscaglia's Restaurant on Jackson Gate Road for a meal of roast turkey, baked chicken, ravioli, taglierini, salad and wine, all for 75¢.

   From left to right are Adrian Cooper, trumpet; Leonard Covello, banjo; Ernie Massei, violin and saxophone; bandleader Art Caviglia, accordian and piano; Ted Shipman, saxophone; band manager Louie Caviglia; and Harley Boseman, saxophone. In the back row are Elmyran Cooper, bass, and Bob Kerr on drums.

A crowd of ten thousand attended a rock concert at the Calaveras County Fairgrounds in Angels Camp in 1978.

Many of the Mother Lode towns hold flea market sales on weekends. This one was held in Jackson during the Dandelion Festival in March 1979. The old I.O.O.F. Hall on Main Street can be seen in the far background.

A crowd on the hillside in Frogtown, just outside Angels Camp, during the Jumping Frog Jubilee.

An orderly crowd of fifty thousand young people heard seventeen hours of rock music at tiny Amador Lake, ten miles west of Jackson on October 4, 1969.

A Mother Lode Bibliography

For additional information about the Mother Lode Country, we suggest the following titles:

Andrews, John R. *Ghost Towns of Amador*. Second edition. Fresno: Valley Publishers, 1978.

Beilharz, Edwin A. and Lopez, Carlos U., editors. *We Were 49ers, Chilean Accounts of the California Gold Rush*. Ward Ritchie Press, 1976. Available through Valley Publishers.

Buckbee, Edna Bryan. *Pioneer Days of Angels Camp*. Angels Camp: *Calaveras Californian*, 1932.

Buckbee, Edna Bryan. *Saga of Old Tuolumne*. New York: The Press of the Pioneers, Inc., 1935.

Chamberlain, Newell D. *The Call of Gold: True Tales on the Gold Road to Yosemite*. Fresno: Valley Publishers, 1972.

Coy, Owen C. *In the Diggings in 'Forty-Nine*. Los Angeles: California State Historical Association, 1948.

Glasscock, Carl Burgess. *A Golden Highway*. Indianapolis: The Bobbs-Merrill Co., 1934.

Jackson, Joseph Henry. *Anybody's Gold, The Story of California Mining Towns*. San Francisco: Chronicle Books, 1970.

Jenkins, Olaf P. *Geologic Guidebook of the Mother Lode*. The Mother Lode Country. Centennial edition. San Francisco: by Olaf P. Jenkins & Others, 1948.

Masri, Allan and Abenheim, Peter. *The Golden Hills of California, A Descriptive Guide to the Mother Lode Counties of the Southern Mines*. Fresno: Valley Publishers, 1979.

Phillips, Catherine Coffin. *Coulterville Chronicle, The Annals of a Mother Lode Mining Town*. Second edition. Fresno: Valley Publishers, 1978.

Sargent, Shirley. *Mariposa County Guidebook*. Yosemite: Flying Spur Press, 1967.

Snow, Horace. *Dear Charlie Letters*. Mariposa: Mariposa County Historical Society, 1979.

Stoddart, Thomas Robertson. *Annals of Tuolumne County*. Second edition. Fresno: Valley Publishers, 1977.

Sunset. *Gold Rush Country; Guide to California's Mother Lode and Northern Mines*, by editors of Sunset books and Sunset Magazine. Menlo Park: Lane Book Co., 1963.

Taylor, Bayard. *Eldorado*. Palo Alto: L. Osborne, 1968.

Weston, Otheto. *Mother Lode Album*. Stanford University Press, 1948.

Wood, Richard Coke. *Calaveras, the Land of Skulls*. Sonora: Mother Lode Press, 1955.

Wood, Richard Coke. *Murphys, Queen of the Sierra*. Angels Camp: *Calaveras Californian*, 1948.

Wood, Richard Coke. *Tales of Old Calaveras*. Angels Camp: *Calaveras Californian*, 1949.

Photo Credits

We acknowledge with thanks the following photographs contributed to this book:

Daffodil Hill by Barbara Cenotto, p. 61.

Photo of rock concert by Rick Turner, *Amador Dispatch*, p. 202.

Photo of crowd at rock concert by Fred Feary, p. 203.

Aerial photo of rock concert by Larry Cenotto, p. 203.

Index

Adams Express, 70
Agua Frio (Aqua Fria), 167
Albany Flat, 89
Ale and Quail Corporation, 77
Alemany, Archbishop, 176
Aleric, Father Henry, 151
Alpine County, 37, 65, 110
Alpine Mine, 25
Alta Telegraph, 17
Altaville, 78
Alvarado, Governor Juan Bautista, 6
Alvarado, Juan, 168
Amador Central Railroad, 39
Amador City, 37, 38, 53, 54
Amador County, 37-64, 65, 67, 154, 200
Amador County Courthouse, 41
Amador County Fairgrounds, 38
Amador County Museum, 40, 43, 60
Amador Hotel, 53
Amador, Jose Maria, 53
Amador Lake, 203
American Camp, 123
American Forest Products Company, 39, 62
American River, 1, 2, 3, 5, 6, 7, 12, 21, 24, 29, 35
American River (tertiary), 2
American River Land and Lumber Company, 35
"An Episode in Fiddletown," 56
Anderson Hotel, 93
Angel, Henry P., 66, 84
Angels Camp, 66, 67, 71, 72, 77, 78, 83, 84, 85, 86, 89, 109, 116, 196, 199, 202
Angels Creek, 84
Angels Hotel, 117
Angels Mine, 85
Angels Wings, 109
App Mine, 122
Aqua Fria, 167
Aragonite, 112
Archie Stevenot Bridge, 159, 166
Argonaut Mine, 37, 39, 59
Armory (Georgetown), 14
Armory Hall (Copperopolis), 94
Armour, Philip, 19
Arrastras, 24
Art Caviglia Night Hawks, 201
Auburn, 21, 66
Auger, Father, 176

Bagby, 170, 187
Baker, Isaac, 76
Balzar House (Georgetown), 14, 22
"Bandarita" (Bandereta), 171, 191
Barnes, "Batch," 109, 113
Bear Valley, 110, 168, 169, 171, 177, 186, 187
Bear Valley Company, 110
Bedbug (Ione), 62
Bedford Park, 14
Beebe Mine, 25
Bell, Glenn W., Jr., 125
Belotti family, 50
Bendix Corporation, 39, 62
Benton, Senator Thomas Hart, 171, 187
Benton Mills, 187
Best steam engine, 87, 96
Biddle Boggs vs. Merced Mining Company, 168
Big Bar, 37, 65
Big Bonanza Mine, 122
Big Cut, 14
Big Oak Flat Road, 155
Big Trees, 77, 107, 111, 179
Big Trees Grove, 197
Big Trees Hotel, 108, 114, 115, 197
Bigler, Henry, 7, 12
Black Bart, 101, 102, 103
Black Bart Inn and Motel, 101
Black Bart Players, 101, 103
Black Creek Lodge, 162
Black Oak Mine, 25
Blair brothers, 16
Blair's Sly Park Sawmill, 28
Blazing Star Mine, 90
Bodega Bay, 6
Bolton, Charles E. (Black Bart), 101
Boosters' Club of Angels Camp, 109
Booth, Edwin, 199
Booth, Junius B., 148
Boseman, Harley, 201
Botellas (bottle) Spring, 37
Bottle Hill, 26
"Boys Will Be Boys," 43
Bradley, Mrs. Owen, 108
Brignole family, 50
Brown, Armsted C., 43
Brown, Professor DeMarcus, 124
Brown House (Jackson), 43
Bruschi, Francisco, 192

Bruschi, Rose, 192
"Buchanan Mike," 130
Buchanan Mine, 130
Buena Vista, 64
Bunnell, Dr. L. H., 169
Burk, Pat, 130
Burney, Sheriff James, 169
Burns Creek, 170
Burns, Robert, 124
Buscuglias Restaurant, 201
Byrnes, P. O., 159

Calaveras, 65-120
Calaveras Big Trees, 111, 112
Calaveras Cement Company, 76
Calaveras Cement Division of the Flintkote Company, 76
Calaveras Chronicle, 70
Calaveras County, 37, 65, 67, 110, 121, 158, 159, 160, 162
Calaveras County High School, 74
Calaveras County Museum, 87
Calaveras Grange, 99
Calaveras Grove, 107, 108
Calaveras Grove Association, 108
Calaveras River, 65, 71
Calaveras River (tertiary), 2
Calaveritas Camp and Creek, 65
Caldor Company, 33, 35
California Department of Recreation, 10, 23
California Pioneer Towns, 25
California Teachers' Association, 141
California Youth Authority, 63
The Call of Gold, 177
Camanche Dam, 100
Camanche Reservoir, 39
Camino, 28
Camp Nine, 160
Campbell, Dexter, 168, 174
Campbell, Mrs. Margaret, 168
Canyon Creek, 26
Carson City, Nevada, 19
Carson Creek, 66
Carson Emigrant Trail, 24
Carson Hill, 66, 67, 89, 159, 166
Carson Hill Gold Mine, 166
Carson, James H., 66
Carson, Kit, 24, 45
Carson Pass, 39

Carson Pass Road, 45
Carter's, 145
Cary House, 16, 19
Cary, William, 16
"Cathedral of the Mother Lode," 63
Cavagnero family, 50
Cave City, 108
Caves, 108, 109, 112
Caviglia, Art, 195, 201
Caviglia, Louie, 201
Cedar Grove, 16
"Celebrated Jumping Frog of Calaveras County," 109, 116, 117
Cement, 76
Central Eureka Mine, 37, 38, 52
Central Hill Hydraulic Mine, 79
Central Pacific Railroad, 50, 64
Chaffee, James, 147
Chamberlain, Jason, 147
Chamberlain, Newell D., 177
Chavanne Mine, 92
Chaw'se, 61
Chinatown (Sonora), 137
Chinese, 74, 97, 137, 140, 149, 150, 151, 171, 192
Chinese Camp, 149, 150, 151, 155
Chio mine, 80
"Church of the 49ers" (St. Andrew's), 141
Churches, 15, 22, 43, 63, 70, 74, 82, 88, 94, 95, 99, 134, 140, 141, 142, 151, 153, 176, 185
"Clamper Wall," 104
Clark, Galen, 179
Clay, 64
Claypipers, 200
Clay's Bar, 99
Clements, 100
Coakley, Judge Thomas, 170
Coal (lignite), 64
Colfax, Vice President Schuyler, 183
College of the Pacific, 124
Coloma, 1, 5, 6, 7, 10, 11, 12, 13, 15, 24, 25
Coloma Catholic Church, 15
Columbia, 104, 122, 123, 124, 140, 141, 142, 143
Columbia and Stanislaus River Water Company, 123
Columbia Historic State Park, 123, 124, 125, 140, 143
Columbia Junior College, 125, 143
Columbia, South America, 5
Columbia Star, 132
Combellack House, 16
Comstock silver strike, 13
Confidence, 148

Confidence Mine, 148
Consolidated Keystone Mine, 37
Constitutional Convention, 121
Cool Garden Valley, 14
Cooper, Adrian, 201
Cooper, Elmyran, 201
Copper, 96, 97
Copper Cove Resort, 162
Copperopolis, 94, 101, 162
Copperopolis Catholic Church, 95
Copperopolis Congregational Church and Community Club, 94
Copperopolis Drum Corps, 196
Corner, The, 95
Cosumnes River, 5, 24, 37
Coulter, George W., 171, 190, 191
Coulter, Mrs. George, 191
Coulterville, 170, 171, 174, 190, 191, 192, 194
Coulterville Concert Brass Band, 196
Coulterville Hotel, 190
County Hospital (San Andreas), 75
Covello, Leonard, 201
Coyote Creek, 66
Coyote Creek Canyon, 109, 113
Crabtree, Lotta, 199
Crocker, Judge, 173
Cuneo, Don, 103
Curtin, M. J., 131

D'Agostini family, 55
D'Agostini Winery, 38, 55
Daffodil Hill, 39, 61
Dambacher, Joseph L., 138
Danaker, C. O. Pine Company, 29, 30, 31
Dane, Ezra, 104
Daulton, Merle and Clay, 170
Davidson, Reverend, 53
Deady, D. C., 16
Death Valley, 167
Dexter, John L., 168, 174
Diamond and Caldor Railway, 32, 35
"Diamond Cascade," 112
Diamond Springs, 24
Dinkelspiel Store, 89
Don Pedro Dam, 155, 157, 158, 165
Donkey engine, 33
Dosch pit, 64
Double Springs, 99
Douglas fir, 27
Douglas Flat, 82
Douglas Flat mine, 80

Douglas Flat School, 82
Dowd, A. T., 107
"Dry Diggings," 13
Drytown, 46
Drytown Town Hall, 200
Duran, Father, 158
Dutch Mine, 122

E. Clampus Vitus, 104, 105
Eagle Cottage, 124
Eagle-Shawmut Mine, 122, 155
East Betts, 67, 104, 122
Ebbetts Pass Road, 108
Education and schools, 40, 82, 124, 125, 140, 141, 143, 151
El Dorado, 5, 24, 25
El Dorado County, 5-36, 37
El Dorado County Museum, 15, 28, 32
El Dorado Lumber Company, 29
El Dorado National Forest, 27
Eleanor, Lake, 154
Emerald, Gus, 86
Empire Mine (Plymouth), 38, 54
Estanislao, 158

Fallon House, 124, 142, 198
Fallon, James, 124, 142
Fallon, Owen, 124, 198
Farrell, Al, 90
"Father of the Forest," 197
Feather River, 7
Ferry crossings, 159, 164
Fiddletown, 38, 56
First Congregational Church, 82
Fishing and water sports, 39, 100, 107, 157, 158, 160
Flintkote Company (Calaveras Cement Division), 76
Folsom, 35
Forrester, Frank, 80
Fort Ross, 6, 9
Fourth Crossing, 71
Frances, Joe, 139
Freezeout (Ione), 62
Fremont, California, 178
Fremont Grant, 167, 169, 178, 187, 188
Fremont, Jessie, 171, 178, 186
Fremont, John C., 45, 168, 169, 171, 178, 186
Fresno, 169
Fricot, Desire, 108
"Frogtown," 109, 202
"Frying Pan Grant," 168
Funk Hill, 101

Gale, George, 107
Galt, 64
Garden Valley, 25
Garrett House (Chinese Camp), 150
"Gem of the Southern Mines" (Columbia), 123, 124
Georgetown, 5, 14, 20, 21, 24, 25, 26
Georgetown Hotel, 21
Georgia Slide, 14, 26
Ghirardelli, Domingo, 170, 183
Gibbons, Stuart, 108
Gillis, Bill, 146, 148
Gillis, James, 146, 148
"Glory Hole," 89
Gold Bug Mine, 14
Gold Discovery Day, 25
Gold, discovery of, 1, 6, 7, 12
Gold Hill, 23
Gold Hill Trail Elementary School, 23
Gold Rush, 1, 2, 121
Golden Chain Council, 159
Golden Chain Highway, 159
Golden Gate Mine, 139
Goodyear, Hal, 105
Gothic Chamber, 108
Gottschalk, Judge Victor, 101, 102
Grand Hotel, 69
Grant, President Ulysses S., 115, 183
Greeley, Horace, 16, 19
Grizzly Flat, 26
Groveland, 147
Growlersburg, 14
Guativita, Lake, 5
Guinn Mine, 71
Gulch Creek, 72
Gunn House Hotel, 133

Haferl, Louie, 139
Hance, John William, 67
Hanford, Captain James, 107
Hangman's Tree, 147
"Hangtown," 13, 19
Hard rock mining, 24, 26, 53, 67, 122, 126, 127
Harte, Bret, 56, 147, 159, 164
Harvard Mine, 122
Hasserler, John, 15
Haynes, A. S., 108
Hemminghoffen and Suesdorff brewery, 68
Hetch Hetchy Project, 154
Hetch Hetchy Reservoir, 154

Hetch Hetchy Valley, 158
Hetty Green (Old Eureka) mine, 52
Highway 4, 82, 110
Highway 26, 97
Highway 49, 24, 25, 38, 39, 46, 53, 54, 56, 62, 65, 71, 89, 154, 155, 159, 165, 166, 169
Highway 50, 14
Highway 88, 39, 45
Highway 108, 125
Highway 120, 126, 152, 154
Highway 140, 169
Hildreth, George, 123
Hildreth, Thaddeus, 123
Hildreth's Diggings, 123
"Historic Sites of El Dorado County" (booklet), 25
Hock Ranch, 7
Holden Chispa Nugget, 132
Holden's Gardens, 132
Holt Steamer, 138
"Home Guard" (Georgetown), 14
Hopkins, Mark, 19
Hornitos, 170, 183, 184, 185
Hornitos Hotel, 183
Hornitos Jail, 185
Huff, William Gordon, 104
Humbug Mine, 122
Hume, Sheriff James, 19
Hunt, Mrs. Sadie, 99
Hydraulic mining, 26, 38, 79, 97, 143

IOOF Hall (Coulterville), 192
IOOF Hall (Georgetown), 14
IOOF Hall (Mokelumne Hill), 70
IOOF Hall (Sonora), 169
Immaculate Conception Catholic Church, 99
Incense cedar, 27
Indian Grinding Rock State Park, 39, 61
Indians, 6, 39, 61, 91, 92, 122, 158, 169, 170
Indians of the Yosemite, 179
Inspiration Point, 181
Institute of Forest Genetics, United States Forest Service, 27
Interpace Corporation, 64
Ione, 39, 62, 63, 64
"Ione City Centenary Church," 63
Ione Methodist Episcopal Church, 63
Ione Valley, 62
Irrigation, 157, 158, 159, 165, 166
Italian Picnic, 50

Jackass Hill, 146, 148
Jackson, 37, 40, 41, 42, 47, 60, 62, 154, 198, 201, 202, 203
Jackson Butte, 40, 41, 68
Jackson, Colonel Alden Apollo Moore, 37, 40, 154
Jackson Elementary School, 40
Jackson Gate, 58
Jackson Gate Road, 43
Jackson Military Band, 197
Jackson's Camp, 154
Jacksonville, 154, 155, 157
Jacksonville Reservoir, 154, 155
James, Colonel George, 121, 126
Jamestown, 121, 122, 126, 127, 162
Jamestown Depot, 127, 200
Japanese American Citizens League of El Dorado County, 23
Jeffery Hotel, 171, 190, 192
Jeffrey pine, 27
"Jimtown," 121
Johnston, Wade, 67
Josephine Mine, 168, 169
Joy, Emmett, 74
Jumping Frog Jubilee, 76, 109, 110, 116, 118, 202
Juniper Mine, 122
Jurassic Sea, 1, 3

Kahn, Dr. Sy, 124
Kaler, Pat, 80
Kearny, General Stephen, 168
Kelsey, 7, 14
Kelsey, Benjamin, 14
Kelsey, Samuel, 14
Kennebec Hill, 142
Kennedy Mine, 37, 38, 39, 57, 58, 59, 60
Kentucky House, 76
Kerr, Bob, 201
Keys, John, 51
Keystone Consolidated Mine, 38, 53, 54
Kirkwood, 39
Knickerbocker Flat, 143
Knight Foundry, 38, 52
Knight, Samuel N., 38, 52
Knight's Ferry, 152, 153, 159, 161
Knight's Ferry Methodist Episcopal Church, 153
Knights water wheel, 38, 52
Knox, Shannon House (Georgetown), 14, 20

La Grange, 157, 158
"La Mariposa," 168
La Meneta, 188

Lake Camanche Village, 99
Lake, Judge Delos, 84
Lake McClure Reservoir, 187
Lapham, W. W., 108
Larkin, Thomas O., 168
Lassagne, Art, 25
Leger, George, 69
Leger Hotel, 68, 69, 70
Le Vaggi family, 50
Liberty, Leo, 195
Life and Adventures of James Marshall The, 7, 11
Lignite (coal), 64
Limestone, 76
Limestone Caves, 108
Lincoln (Union) Mine, 37, 50
Lind, Dr. John, 97
Lind, Jenny, 97
Lippincott, Oliver, 194
Lipton, Thomas A., 115
Lititz, Pennsylvania, 7
"Little White House," 171, 186
Long Barn, 148
Lotus, 25
Louisiana House (National Hotel) (Jackson), 42
Lumber industry
  Amador County, 39, 62
  Calaveras County, 39, 71, 86, 100
  El Dorado County, 27-36, 39
  Tuolumne County, 138, 144, 145

M cDougal, Governor John, 169
McIntyre, Clara, 51
McKinley, President William, 126
McLaughlin family, 61
MacNider, William, 76
Mac's Place, 45
Magnolia Saloon and Museum, 171, 193
Malakoff Diggings, 2
Malatesta family, 50
Man-Chow temple (Chinese Camp), 150
Manuel Lumber Company, 86, 87
Mariposa, 66, 167, 169, 175, 177, 186
Mariposa Battalion, 169, 181
Mariposa Chronicle, 168, 174
Mariposa County, 13, 24, 121, 167
Mariposa County Courthouse, 167, 168, 172, 173
Mariposa County Guidebook, 170
Mariposa County Historical Society, 170, 182

Mariposa County Jail, 176
Mariposa Gazette, 168, 174
Mariposa Grant, 167, 168, 171, 177
Mariposa Grove, 179
Mariposa History Center, 170, 182
Mariposa Mine, 169
Mariposa Museum, 175
Mariposa quartz gold mine, 24
Marshall Gold Discovery State Historical Park, 13
Marshall, James W., 1, 6, 7, 10, 11, 12, 14, 27
Martell, 39, 62
Martin, Reverend, 134
Martinez Mine, 24
Mary Harrison Mine, 171, 194
Masonic Cave, 48
Masonic Lodge (Sonora), 130
Massei, Ernie, 201
Maxwell, George, 171
Mead, Stephen, 159, 164
Meador home (Copperopolis), 94
Melones Dam, 160, 166
Melones Bridge, 200
Melones, 66, 159, 160, 164, 166
Mein, William, 76
Merced, 169
Merced Mining Company, 168
Merced River, 2, 187
Mercer, Walter J., 108
Mercer's Caverns, 108, 109, 112
Metropolitan Hotel (San Andreas),
Mexicans, 88, 97, 121, 171, 184, 190
Michelson, Dr. Albert, 82
Micheltorena, Governor, 168
Michigan-California Lumber Co., 29, 30
Middle Bar, 37, 65
Milton, 77, 96, 97
Miner's Union Hall (Chinese Camp), 150
Minister's Gulch, 53
Mirror Lake, 180
"Mr. Mother Lode" (Archie Stevenot), 159
Mitchler (Sperry) Hotel, 77
Mitchel, Cliff, 162
Miwok Indians, 61, 91, 122, 158, 169
Moaning Cave, 108, 109
Modesto Irrigation District, 157, 165
Modesto Junior College District, 124
Moffat and Company, 171
Moffatt, John L., 189
Mokelumne Hill, 65, 66, 68, 69, 71, 104

Mokelumne Hill Community Church, 70
Mokelumne Hill IOOF Hall, 70
Mokelumne River, 2, 37, 39, 65, 68
Monk, Hank (The Whip), 19
Mono County, 65
Monterey *Californian*, 132
Monteverde family, 50
Montez, Lola, 199
Montezuma Mine, 24
Moore, Archie, 197
Moraga, Gabriel, 65, 167
Morgan, Colonel A., 67
Morgan Mine, 67
Mormon Battalion, 6, 12
Mormon Camp, 148
Morris brothers, 150
"Mother of Counties" (Mariposa), 167
"Mother of the Forest," 107, 111
Mother Lode, 1, 3, 5, 13, 37
Mt. Bullion, 169, 170, 171, 188
Mt. Bullion School, 171
Mt. Ophir, 170, 171
Mount Ophir Mill and Mint, 171, 189
Mt. Reba Corporation, 110
Mt. Reba Ski Resort, 110, 120
Mountain Pass Road, 162
Mountain Ranch, 93, 108
"Mud Springs," 24
Muir, John, 180, 197
Murietta, Joaquin, 170, 184
Murphy, Dan, 66
Murphy, John, 66, 78
Murphys, 76, 77, 78, 79, 80, 81, 82, 83, 86, 87, 91, 101, 104, 105, 108, 109
Murphys Camp, 66, 108
Murphys Diggings, 78, 107
Murphys Flats, 78
Murphys Hotel, 77, 78
Murphys Independent Cornet Band, 195, 197
Murphys New Diggings, 66
Murphys Old Diggings, 88
Murphys School and Community Center, 82
Murphys South Ditch flume, 80
Museums and historical sites, 2, 5, 7, 9, 10, 11, 12, 13, 15, 16, 28, 32, 38, 40, 48, 55, 60, 73, 87, 93, 103, 104, 123, 170, 171, 175, 182, 193

N ahl, Charles, 10
Napoleon copper mine, 96
Nashville, 24
Nashville Mine, 25

National Hotel (Louisiana House, Jackson), 42
Native Daughters of the Golden West, 44
Native Sons of the Golden West, 7, 12, 74, 88
Native Sons Hall (Murphys), 101
Natural Bridges, 109
Nava, Joe, 86
Neocene rivers, 2
Nevada, 13
Nevada City, 2
Nevada County, 2, 24
New Melones Dam, 160, 166
New Melones Reservoir, 159, 166, 200
New York Volunteers, 38, 47, 66
North Ditch flume, 78, 79
North Grove (Calaveras Big Trees), 108
North Grove Road, 108
Northern Mariposa County History Center, 193
Nueva Helvetia ("New Switzerland") (New Helvetia), 6, 27

Oakdale Irrigation District, 159, 163, 166
Oakhurst, 169
O'Byrnes Ferry, 159, 162, 163, 164
Okei, 23
"Old Abe" (cannon), 49
Old Covered Bridge, 159
Old Eureka (Hetty Green) mine, 52
Old Tecumseh, 91
"Old Timers Museum," 81, 104, 105
Oleta, 38, 56
Olmstead, Mrs., 22
Oregon Bar, 37, 65
Oregon Canyon, 14
Organ Loft, 109
Oro y Plata Mine, 108
Orvis and Associates, 110
Orvis, C. Bruce, 110
Oso House Hotel, 171, 186
"Outcasts of Poker Flat," 159, 164
Overland Route, 13
Owens-Illinois glass plant, 64

Pacific Mine, 54
Paleozoic era, 1
Papeete, 123
Pardee Dam and Reservoir, 100
Pardee Reservoir, 39
Parrots Ferry, 159, 160

Parrots Ferry Road, 109
Parsons, George, 7, 11
Patten and Springer Orchestra, 195, 201
Patten, Bob, 201
Pepper, Vernon, 171
Pep's Museum, 193
Perry, John, 108
Perry, Viola, 70
Perry's Murphys Hotel, 108
Peter's Bakery, 70
Peters, Charley, 4
Piety Hill, 122
Pilot Hill, 14
Pine Grove, 45, 61
Pine Grove House (hotel), 45
Pine Tree Mine, 168, 169
Pino Grande, 28, 29, 35
Pioneer Hall (Jackson), 44
Piper Playhouse, 200
Phipps, George, 14
Placer County, 24, 67
Placer mining, 24, 26, 37, 53, 65, 66, 121, 122, 170
Placerville, 5, 13, 14, 15, 16, 17, 18, 19, 21, 24, 26, 27, 28, 29
Placerville-Carson Valley Road, 13
Placerville Tree Nursery, 27
Pleistocene rivers, 2
Plumas County, 169
Plymouth, 37, 46, 54, 55, 56
Plymouth Consolidated Mine, 37, 38, 54
Poker Flat, 164
Pokerville, 54
Pollardville, 93
Ponderosa pine, 27, 34
Pony Express, 13, 16
Pony Express Trail, 16
Priest, W. C., 155
Priest's Grade, 155
Priest's Hotel, 155
Preston School of Industry, 63
Princeton, 171, 188
Princeton Mine, 169, 188
Princeton Saloon, 171

Quartz mining, 24, 25, 37, 53, 67, 85, 92, 122, 126, 167, 168, 177
Quartzburg, 170
Quartzville, 25
"Queen of the Southern Mines" (Sonora), 122

Raggio brothers, 87
Raggio, Charley, 195

Railtown, 127
Ralston, Julia, 133
Rasmussen, Dennis, 110
Ratto Theater, 198
Rawhide Mine, 122
Red Bird Mine, 24
Reichling, Lily, 44
Reservoirs
  Camanche, 39, 99
  Hetch Hetchy, 154
  Jacksonville, 154, 155
  Lake McClure, 187
  New Melones, 159, 160, 166, 200
  Pardee, 39, 100
  San Diego, 125
  Tulloch, 159, 162, 163
Rhodes, Harvey, 125
Rich Bar, 65
Rich Gulch, 71
Riley, Governor Bennett, 121
Rio de las Calaveras ("River of Skulls"), 65
Robinson, John, 159, 164
Rogers, Will, 43
Romaggi, James, 89
Roosevelt, President Theodore, 171
Rosenblohm's Store, 150
Rossi, Angelo J., 49
Rothschild's, 115

Sackett, Ed and Vi, 171
Sacramento, 8, 21, 49, 90, 104, 195
Sacramento River, 6
Saint Andrew's Catholic Church, 74
St. Andrew's Presbyterian Church, 141
Saint Anne's Catholic Church, 142
St. Catherine's Catholic Church, 185
Saint Francis Xavier Church, 151
St. George Hotel (Volcano), 47
Saint Ignacio, 147
St. James Catholic Church (Georgetown), 22
St. James Episcopal Church, 134
St. Joseph's Catholic Church, 176
St. Patrick's Catholic Church, 43
St. Sava's Serbian Orthodox Church, 43
Sam-Yap, 149
San Andreas, 67, 69, 71, 72, 73, 74, 76, 80, 93, 98, 101, 102, 103
San Andreas Joss House, 74
Sand, 64
San Domingo Road, 108

San Diego Reservoir, 125
San Francisco, 104, 113, 158, 165
San Francisco Foundry, 12
San Joaquin and Sierra Nevada Railroad, 98
San Joaquin Delta College, 102
San Joaquin District, 121
San Joaquin River, 66, 158
Sargent, Shirley, 170
Savage, Major James, 169, 181
Sawmill Flat, 122
Second Garrotte, 147
Senter building (Murphys), 81
Senter, Riley Store (P. L. Traver General Store), 81
"Sentinels" (Big Trees), 114
Sequoia Gigantea (Sierra Redwoods), 107, 111
Shawmut Grade, 155
Shawmut Mine Chlorine Works, 155
Sheep Ranch, 91, 92
Shenandoah Valley, 38, 54
Shingletown, 34
Shipman, Ted, 201
Sierra County, 24
Sierra Gold Belt, 24
Sierra Nevada, 1, 2, 3, 27
Sierra Nevada House, 15
Sierra Railway Depot, 127, 131
Sierra Redwoods (*Sequoia Gigantea*), 107
Skiing, 39, 110, 120
Sly Park, 28
Smith, Judge James Alexander, 75
Snider Lumber Company, 100
Soldier's Gulch, 47
Sonora, 122, 123, 124, 125, 128, 129, 130, 131, 132, 135, 138, 140, 154, 159, 166, 196
Sonora City Hotel, 132
Sonora City Jail, 135
Sonora Herald, 132
Sonora Independent Hose Company #2, 137
Sonora Methodist Episcopal Church, 134
Sonora Pass, 125, 129
Sonora Theater, 131
Sonora Race Track, 136
Sonorian Camp, 121, 123
Soulsby Mine, 122, 127
South Eureka mine, 52
South Grove (Calaveras Big Trees), 108
South San Joaquin Irrigation District, 159, 163, 166
Southern Mines, 24, 65, 83, 121, 122, 159, 170, 183, 195
Southern Pacific Railroad, 39, 96, 98

Sperry (Mitchler) Hotel, 77, 101
Sperry, James L., 108, 115
Sperry, Mehitable, 115
Sportsmans Hall, 16
Springer, Eli, 201
Squabbletown, 122
Stamp mills, 24, 50, 54, 58, 182, 187
Standard, 144
Standard Lumber Company, 138
Stanford, Leland, 50
Stanislaus County, 158
Stanislaus National Forest, 110
Stanislaus River, 2, 65, 66, 67, 80, 107, 108, 121, 123, 152, 157, 158, 159, 161, 163, 164, 200
State Parks System, 108, 123
Steinmatz, D. H., 138
Stevenot, Archie, 105, 159, 166
Stevenson, Colonel Jonathan, 38, 47, 66
Stickle's Theater, 199
Stockton, 19, 66, 71, 83, 89, 93, 104, 159, 177, 194, 195
Stockton and Copperopolis Railroad, 96
Stockton, Bill, 86
Stockton Creek, 177
Stockton Mining Co., 19, 66
Stockton-Murphys Stage, 73
Stockton, Commodore Robert F., 177
Stockton Times, 132
Studebaker Building (Placerville), 18
Studebaker, John M., 18, 19
Sugar pine, 27
Sutter Creek, 37, 38, 50, 52
Sutter Creek Inn, 51
Sutter, Captain John, 6, 7, 8, 9, 12, 27, 38, 50
Sutters Fort, 6, 9, 27
Sutter's Mill, 10, 11
Sweeney Mine, 122
Swerer Store, 148

T able Mountain, 122, 164
Taylor, Bayard, 5
Tehachapi Mountains, 167
Temperance, 45
Tenaya, Chief, 169
Tennessee-Nashville Mine, 24
"Tennessee's Partner," 147
Tertiary rivers, 2
Thatcher Mill, 34
"The Luck of Roaring Camp," 159
Theater, 14, 124, 131, 140, 195, 198, 199, 200

Thompson, John A. "Snowshoe," 19
Thorn, Thomas, 170
Thorne, Ben, 103
Toll's Express, 89
Trabucco, Judge J. J., 174
Trabucco's Store, 171
Tracy, 64
Traver, P. L. General Store (Riley Senter Store), 81
Tri-Dam Project, 159, 162, 163
Troy Gold Industries, Ltd., 90
True, Edward, 86
Tuleburgh, 66
Tulloch, Charles, 163
Tulloch Grist Mill, 153
Tulloch Reservoir, 159, 162, 163
Tuolumne, 125, 144, 145
Tuolumne City Intelligencer, 132
Tuolumne County, 121, 166, 167
Tuolumne County Historical Society Museum, 135
Tuolumne County Hospital (Sonora), 131, 133
Tuolumne County Water Company, 123
Tuolumne River, 2, 121, 154, 155, 157, 158, 165
Tuolumne River (tertiary), 2
Tuolumne Square Shopping Center, 145
Tourist attractions, 24, 38, 39, 107, 108, 109, 110, 120, 124, 125, 158, 160, 169, 181
Turlock Irrigation District, 157, 165
Tuttle, Judge Anson A. H., 148
Tuttletown, 148
Tuttletown Hotel, 148
Twain Harte, 148
Twain, Mark, 109, 115, 116, 117, 146, 148

U hlinger family, 55
Uhlinger, Jacob, 55
Utica Mine, 83, 84
Utica Mine Brass Band, 196
Union Hotel (Copperopolis), 95
Union (Lincoln) Mine, 24, 37, 50, 52
Union Water Company, 78, 107
United States Forest Service, 39
 Institute of Forest Genetics, 27
University of the Pacific, 124, 198

V allecito, 66, 76, 79, 80, 82, 88, 89, 108, 109, 113, 160

Vallecito Community Church, 88
Vallejo, General Mariano, 122, 158
Valley Springs, 98, 100, 201
Victoria Hotel (Sonora), 132
Vinton, 169
Virginia City, Nevada, 13, 104
Volcano, 37, 38, 39, 47, 48, 49, 61
Volcano Blues, 49

Wakamatsu Tea and Silk Colony, 23
Walker, Joseph, 169
"Wall of Comparative Ovations" (Murphys), 104, 105
Wallace, 100
Warren, Governor Earl, 123, 140
Washington tree, 111
Waterford, 194
Waters, James, 67
Watson, Polly, 195
"Wawona" Tree and Tunnel, 179, 181
Webber Creek, 19, 66

Weber, Captain Charles, 19, 66
Weimer, Ray, 118
Weir Brewery, 150
Wells, E. G., 183
Wells Fargo, 16, 17, 70, 78, 81, 89, 101, 150, 171, 183, 193
Weisbad, Louis, 67
West Belts, 67, 122
Western Pine Association Tree Farms, 27
West Point, 90
Westside and Cherry Valley Railway Park, 125
Westside Lumber Company, 125, 144
Wheat, Alexander, 99
Wheat, Carl, 104
"Whip," The (Hank Monk), 19
"Whistling Billy" (locomotive) 171, 194
White fir, 27
Whiteside, Joseph, 22
Whitney, Mount, 167
Whitsell, Leon, 104
Williams, Price, 80
Willow Hotel, 126

Wineries, 38, 55
Wood, Coke, 81, 105
Wood, Ethelyn, 81
Wood, Reverend James, 121
Wood, Percy, 159
Woods' Creek, 121, 126

Yankee Hill, 122
Yan-Woo, 149
"Ye Old Moke-Hill Bakery" (Peters), 70
Yosemite Falls, 179, 181
Yosemite Junior College District, 124
Yosemite National Park, 150, 155, 157, 169, 181
Yosemite Valley, 169, 170, 180, 181, 194
Yuba River, 2
Yuba River (tertiary) 2

Zamorano, Argustin, 132
Zumwalt, Joe, 104

211